Automotive Development Processes

Julian Weber

Automotive Development Processes

Processes for Successful Customer Oriented
Vehicle Development

 Springer

Dr. -Ing. Julian Weber, Adjunct Associate Professor (Clemson University)
Strategy Manager
Product Strategy Vehicles
BMW Group
80788 Munich
Germany
julian.weber@bmw.de

ISBN 978-3-642-01252-5   e-ISBN 978-3-642-01253-2
DOI 10.1007/978-3-642-01253-2
Springer Dordrecht Heidelberg London New York

Library of Congress Control Number: 2009928427

Cover design: eStudio Calamar S.L., Heidelberg

Printed on acid-free paper

Springer is part of Springer Science+Business Media (www.springer.com)

Foreword

The global crisis the automotive industry has slipped into over the second half of 2008 has set a fierce spotlight not only on which cars are the right ones to bring to the market but also on how these cars are developed. Be it OEMs developing new models, suppliers integerating themselves deeper into the development processes of different OEMs, analysts estimating economical risks and opportunities of automotive investments, or even governments creating and evaluating scenarios for financial aid for suffering automotive companies: At the end of the day, it is absolutely indispensable to comprehensively understand the processes of automotive development – the core subject of this book.

Let's face it: More than a century after Carl Benz, Wilhelm Maybach and Gottlieb Daimler developed and produced their first motor vehicles, the overall concept of passenger cars has not changed much. Even though components have been considerably optimized since then, motor cars in the 21st century are still driven by combustion engines that transmit their propulsive power to the road surface via gearboxes, transmission shafts and wheels, which together with spring-damper units allow driving stability and ride comfort. Vehicles are still navigated by means of a steering wheel that turns the front wheels, and the required control elements are still located on a dashboard in front of the driver who operates the car sitting in a seat.

However, what has changed dramatically are processes involved in vehicle development. What used to be solely the work of one brilliant engineer over several years is achieved today by a highly interlaced co-operative network of specialists coming from a variety of disciplines. The process of vehicle development has become a complex interplay of decentralized sub-processes which are steered on a relatively high level. Even though this has been the dream of automotive development managers for years, there is no such thing as a completely detailed process model. On one hand, if there were one, it would be out-of-date the day after it was completed. On the other hand, on the operational level, real vehicle development "happens" to a certain extent according to individual experience, preference, and current necessities, rather than following a meticulously detailed plan. Even at the most efficient carmakers in the world, it is, to a surprisingly high extent, an ad-hoc process. After all, automotive development is about people.

It is that twofold challenge, to both technically integrate separate components to create a complete vehicle, and at the same time to orchestrate the cooperation of thousands of people from different companies and different professional, cultural and social backgrounds, which makes automotive development so challenging and fascinating. The graduate course in Automotive Development Processes which I have had the opportunity to teach at Clemson University's International Campus for Automotive Research (ICAR), and which is the basis for this book, focuses on two topics: first, the realization of customer relevant vehicle characteristics, and second on the people involved: their personal objectives, their way of thinking and their interaction. I hope this book reflects and summarizes all of the fruitful discussions I have had with automotive experts from the most diverse areas, as well as my own personal experience gained over many years in the field of product development.

In this sense, this book is a personal report rather than a manual for vehicle development. It immerses the reader in the wide range of automotive development processes: from project milestones down to virtual collision checking; from product strategy to production and service integration; from agility to sustainability; and from E/E architecture to embedded software. My intention is to make the reader familiar with the entirety of what people really do in contemporary automotive development, rather than to discuss technical details in-depth. For example, for a passive safety engineer, the chapter on passive safety might only reflect his or her basic knowledge, but by reading through other chapters he or she can gain insight into the processes and the driving forces of neighboring departments and eventually get a better understanding of his or her job in the global context of automotive development.

Compared to other publications on automotive development, the approach followed in this book reflects a customer's rather than an engineer's point of view. It is my strong conviction that in automotive development, customer relevant vehicle characteristics must steer the concept and components, not the other way round. If eventually functions and properties such as agility, passive safety, cabin comfort or even cost suit the customers' requirements, the underlying technical solutions, such as the chassis concept, are of minor importance.

I hope that this book will help managers, specialists, consultants, analysts, students or anyone else interested in the field of automotive development, to better understand the overall process of motor vehicle development; and to recognize the technical and human relationships, dependencies and conflicts between the different sub-processes and the people involved. And lastly, I hope to share my fascination for this exciting profession.

Munich   Julian Weber

Acknowledgements

After teaching a graduate course in Automotive Development Processes at Clemson University's International Campus for Automotive Research (ICAR) for two years, it was the faculty at ICAR that gave me the igniting spark for this book. I would like to thank Dr. John Ziegert for his ongoing support over the past years, and especially for reviewing the book both in terms of content and language. I would also like to thank Dr. Imtiaz-ul Haque, Dr. Thomas Kurfess, and Dr. Georges Fadel for their continuous involvement and encouragement.

A comprehensive characterization of automotive development processes is not base upon a single person's expertise. Numerous contributions from industry experts provide the intellectual foundation which made this book possible. The following people have especially shared their vast knowledge: Rainer Andres, Hans Baldauf, Dr. Jens Bartenwerfer, Dr. Jochen Böhm, Dr. Andreas Goubeau, Dr. Michael Haneberg, Dr. Florence Hausen-Mabilon, Dr. Dieter Hennecke, Martin Hofer, Reinhard Hoock, Dr. Todd Hubing, Gerd Huppmann, Benoît Jacob, Thomas King, Klaus Kompass, Carl-August von Kospoth, Wolfgang Kühn, Johannes Meisenzahl, Reinhard Mühlbauer, Dr. Herbert Negele, Dr. Ulf Osmers, Andreas von Panajott, Dr. Steffen Pankoke, Michael Pfunder, Dietger Pollehn, Kristina Posse, Dr. Friedrich Rabenstein, Dr. Günter Reichart, Tim Rhyne, Dr. Erich Sagan, Harald Schäffler, Axel Schröder, Dr. Verena Schuler, Hans Schwager, Dr. Rudolf Stauber, Wolfgang Thiel, Hoang Phuong Than-Trong, Dr. Gerhard Thoma, Volkmar Tischer, Dr. Ulrich Veh, Erich Wald, Cornelia Würbser, Hannes Ziesler, Andreas Zimmermann.

In addition to individual contributions, many companies and institutions have supported through allowing the usage of proprietary documents. Foremost, I would like to thank the BMW Group of Munich, Germany, for their strong cooperation and permission to publish relevant development material. Other intellectual property documents are reprinted with the kind permission of the following companies and institutions: Allgemeiner Deutscher Automobil Club (ADAC), Association for the Advancement of Automotive Medicine (AAAM), Autoliv, AUTOSAR GbR, AZT Automotive GmbH, California Air Resources Board (CARB), Carnegie Mellon University Software Engineering Institute, Dr. Ing. h.c. F. Porsche Aktiengesellschaft, dSPACE, Inc., EFQM, Environmental Protection Agency (EPA), Euro NCAP, First Technology Safety Systems (FTSS), FORD, Gesamtverband der Deutschen Versicherungswirtschaft e.V. (GDV), Group Lotus Plc., Human Solutions GmbH, Interbrand Zintzmeyer & Lux AG, International Organization for Standardization (ISO), International TechneGroup Incorporated (ITI), J.D. Power and Associates, Lamborghini SA, MAGNA Steyr, Original Equipment Suppliers Association (OESA), Pierburg Instruments, Relex Software Corporation, Renault Deutschland AG, Securmark AG, Shell Deutschland Oil GmbH, Tesla Motors, Inc., The Motor Insurance Repair Research Centre (MIRRC), Toyota Deutschland

GmbH, United Nations Economic Commission for Europe (UNECE), VDI Verein Deutscher Ingenieure e. V., Verband der Automobilindustrie e.V. (VDA), Volkswagen AG, Wölfel Beratende Ingenieure GmbH & Co. KG, ZF Lenksysteme GmbH.

Additional Remarks

The tables and text in Sect. 5.2.8, pages 74 through 77 (the "Adapted Material") have been created from the Technical Report, CMMI® for Development, Version 1.2, CMU/SEI-2006-TR-008, (c) 2006 Carnegie Mellon University and special permission to create and use the Adapted Material has been granted by the Software Engineering Institute of Carnegie Mellon University. CMMI and Capability Maturity Model are registered trademarks of Carnegie Mellon University. Any Carnegie Mellon University and Software Engineering Institute material contained herein is furnished on an "as-is" basis. Carnegie Mellon University makes no warranties of any kind, either expressed or implied, as to any matter including, but not limited to, warranty of fitness for purpose or merchantability, exclusivity, or results obtained from use of the material. Carnegie Mellon University does not make any warranty of any kind with respect to freedom from Patent, Trademark, or Copyright infringement. The Software Engineering Institute and Carnegie Mellon University do not directly or indirectly endorse nor have they reviewed the contents of this book.

Figure 6.12 taken from ISO 9001:2000 *Quality Management Systems – Requirements* is reproduced with the permission of the International Organization for Standardization (ISO). This standard can be obtained from any ISO member and from the Web site of the ISO Central Secretariat at the following address: www.iso.org. Copyright remains with ISO.

EFQM Excellence Model (Fig. 6.13) is the copyright and trademark of EFQM.

The picture of the Bentley Arnage interior shown in Fig. 7.6 has been taken by Jim Callaghan.

Tables 7.7 and 7.8 are reprinted with permission from the Association for the Advancement of Automotive Medicine (c) AAAM.

Contents

Chapter 1
Vehicle Development Projects – An Overview

Abstract   Vehicle development projects may range from a solitary model to a comprehensive model line with multiple variants and derivates, or from a simple facelift to a complete redesign. In any case, development follows a well-planned *product evolution process*, the so-called PEP. The PEP is the core process that transforms the strategic vision of a car into the reality of the first customer vehicle.

1.1 Categories of Vehicle Development Projects

The industrial development of motorized vehicles is usually organized in projects [1]. Such vehicle development projects vary greatly in terms of required technical content, financial effort, and length of time. The main parameters that drive the required effort are:

- Design level
- Design content
- Innovation level
- Number of options

1.1.1 Design Level

The design level of a vehicle development project describes where the project starts and thus determines the required effort. In order from high to low effort, the usual design levels are:

- *Complete redesign.* Starting from scratch, both concept and components are newly designed. Standard and carry-over parts are used only in non-visible areas. As an industry-wide rule, the life cycle of a car is seven years, so models are typically redesigned every seven years. Redesigns require the biggest effort for planning, designing and testing and thus are the most costly development projects.
- *Derivative design.* Redesigning a car based on an existing platform and system architecture (see Sect. 5.2.4). While parts and systems are reused to minimize development and production costs, the customer should - at least at first

J. Weber, *Automotive Development Processes*, DOI 10.1007/978-3-642-01253-2_1,

sight – not be aware of any commonality between the base vehicle and the derivative.[1]

- *Variant design.* In contrast to derivatives, *variants* visibly build a family of cars (see the variants of the BMW 3 Series in Fig. 1.2). Usually, alternative body types such as coupe, wagon or convertible are derived from a sedan. In addition to platform and architecture, parts of the body and exterior trim as well as interior components are carried over from the base vehicle. The effort required for designing a variant largely depends on whether the variant was already planned as a member of a model line during the design of the base vehicle (see Sect. 1.2.2).
- *Model updates* are minor design changes intended to raise the value (and thus the retail price) of a model after the first half of its life cycle. Usually, these changes include exterior trim parts (the reason why a model update is also referred to as a *facelift*), interior trim or new colors and options. The target is, to achieve a newer and fresher look-and-feel at the lowest possible development cost.
- A *model year* project summarizes changes required for cost or quality reasons. These changes are typically collected over a year and brought into production after the summer production shutdown. This allows minimal interruption of series production and the possibility to change production equipment accordingly if required.

1.1.2 Design Content

Another parameter that steers the complexity of a development project is the required design content. The more and the more complex functions the new vehicle offers to the customer, the more effort has to be put into design, evaluation and validation. Relative to the base vehicle, the usual indicators for design content include:

- Number of parts
- Number of electronic control units (ECUs)
- Number of lines of vehicle software code

1.1.3 Innovation Level

While technical innovation is one of the main factors that make a vehicle attractive to potential customers, their development increases not only design work, but

---

[1] As an example of a derivative project, the body and interior of the current BMW X3 were all newly designed by Magna Steyr, re-using most of the drivetrain, chassis and lower body parts of the existing 4×4 BMW 3 Series.

especially testing effort on both the component and vehicle level. As no knowledge based on past is available, systems must be evaluated broadly. A higher number of problems can be expected that have to be solved later during the development process.

An example is the front body structure of the current BMW 5 Series. In the previous model, the front body was a pure steel design. Stamped parts of different steel grades were spot-welded together – a well known process with lots of data available describing operational strength, corrosion behavior, crash worthiness, aging characteristics etc. Evaluation of this design is more or less a standard procedure. The current 5 Series however is equipped with a front body structure that is composed of steel parts, aluminum parts, cast parts and plastic parts which are spot-welded, laser-welded, glued, or riveted together. This highly innovative solution required extensive – and thus costly – testing to ensure safety and functionality in every possible driving situation and durability over the whole vehicle lifetime.

1.1.4 Options and Country Versions

The major driver for complexity and thus for evaluation effort is the number of options offered in a vehicle. Premium brands typically offer the broadest set of freely combinable features to enable the customer to configure the car exactly to his or her needs and desires. While additional options might contribute to customer satisfaction and trigger the decision to purchase, they exponentially increase the required testing effort. For the new 2009 BMW 7 Series e.g., over 200 different options can be selected and combined – in addition to 12 different exterior colors and 12 different trim colors. This creates – theoretically – 3.5 E30 possible vehicle configurations, each of which should be geometrically and functionally evaluated to ensure 100% reliability.

The first approach to reduce complexity from options is to bundle them. If, for example, three levels of stereo systems (none, low, high) and three levels of navigation systems can be selected, then there are nine design combinations for these features. As the take rate for combinations of high stereo with none navigation or no stereo with high navigation is normally very low, it might make sense to offer only three stereo/navigation bundles: none/none, low/low, high/high – thus saving evaluation effort for six combinations.

A strategy followed by some Japanese *original equipment manufacturers* (OEMs) is to evaluate only the most frequently selected 20% of possible combinations, which typically represents over 95% of the vehicles ordered. If a customer selects a configuration that has not been evaluated before, it is then evaluated immediately – leading to a slight delay in delivery time for this vehicle. With this Pareto-approach, the full scale of independently combinable options can be offered while evaluation effort is greatly reduced. For a few customers however who order rare vehicle configurations, the wait for their vehicle's delivery can be very long then, because part of the vehicle's development is only started after their order is placed.

In addition to the options, legislation in different markets requires country specific versions. Acceptable emission levels, crash standards, and safety features differ especially between Europe, the U.S. and Japan (see Sect. 7.1). The country version that requires the biggest modification is the right hand drive which is mandatory e.g. in Great Britain, South Africa and Japan and induces variant parts for body, chassis, steering, dashboard, interior trim, harness etc. OEMs selling internationally usually offer three versions of their base cars: *Europe*, *U.S.* and *Right Hand Drive*.

1.2 Platforms and Model Lines

An established approach to develop more cars faster and at lower cost is the use of components in multiple different vehicles or platforms. Sharing standardized building blocks e.g. for electronic components, chassis systems or engines over several variants, model lines, brands or even OEMs lead to:

- Reduced costs and time required for component design and evaluation
- Reduced demand for tooling and equipment, including reduced costs and time required for design, manufacturing, handling and maintenance
- Increased vehicle quality through usage of mature and well-known components

Application of building blocks relates mainly to non-visible or non-differentiating areas of the vehicle. Two common strategic approaches for concentrated re-use of components are platforms and model lines.

1.2.1 Platforms

A platform[2] is a shared set of components common to a number of different vehicles which may also belong to different brands. The target is, to get maximum differentiation between the cars of one platform while sharing a maximum of parts. Most niche vehicle projects such as roadsters or sports utility vehicles (SUVs) would not be economical without reusing an existing vehicle architecture [2].

---

[2] Originally, a platform was a chassis that was engineered for one and then reused for another car. For example the chassis frame of the Volkswagen Beetle was reused for the Volkswagen Karmann Ghia in 1954.

Probably the most consistent platform[3] strategy today is followed by the Volkswagen Group: The so-called Golf-platform is shared by 4 different brands and a total of 13 different models[4] and includes most parts of the powertrain, steering and suspension as well as parts of the lower body and interior trim (see Fig. 1.1). Differentiation takes place by exterior body and interior trim parts. Interface parts that connect the platform to the model-specific body are customized. Some parts are only differentiated by the attached brand labels (e.g. steering wheel cover).

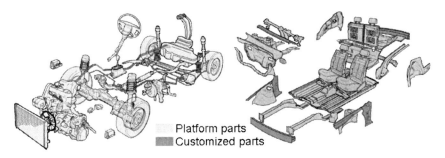

Fig. 1.1 Volkswagen Golf platform PQ34 (Source: Volkswagen)

1.2.2 Model Lines

While the commonality of vehicles sharing one common platform should not be immediately visible, vehicles belonging to one model line do also share exterior and interior parts, which makes them not only technically but also visibly members of one family. This resemblance can be seen e.g. among the variants of the BMW 3 Series model line (see Fig. 1.2).

In the past, model lines were created by first independently designing a base vehicle and then deriving variants. During the variant design, components which had been agreed as communal often had to be changed later on, spoiling parts of the intended savings in development costs. However, the full financial potential of a model line can only be tapped if all vehicles belonging to it and their shared parts and components are planned in advance. Basically this means, that concepts for all member vehicles must be ready and consistent before the first car to be launched goes into series development. A shared lower body structure for the basis sedan e.g. must be designed and proven feasible for coupe, wagon, convertible, 4×4 etc.

---

[3] Engine, gear box, engine mount, front axle, steering gear, steering column, gear control, pedal system, rear axle, brake system, fuel system, exhaust, wheels, tires, front body structure, bulk head, lower body structure, rear body structure, seat frames, platform harness.

[4] VW Golf hatch, Golf wagon, Bora sedan, Bora wagon, New Beetle, New Beetle convertible; Audi A3, TT, TT roadster; Skoda Oktavia, Oktavia wagon; Seat Toledo, Leon.

even if those variants will be produced years after the basis. Equally, a consistent production strategy (which vehicle will be built at which plant) must exist at the same time.

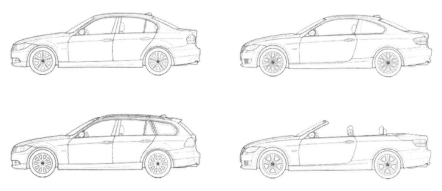

Fig. 1.2 BMW 3 Series family (Source: BMW)

1.2.3 Side Effects / Restrictions

While a consistent platform approach definitely offers advantages regarding development cost and time, it can negatively influence complete vehicle characteristics. If complete vehicle integration measures – e.g. to optimize the vehicle's dynamic driving behavior – may not change the platform components, optimization potential is limited.

From a complete vehicle point of view, a fine differentiation between true carry-over-parts (such as axle links or rims) and parts which should be tunable (such as dampers, engine mounts or tires) is the better solution. This more sophisticated approach requires in-depth experience regarding which parts and properties add up to which complete vehicle characteristic (such as driving behavior, cabin comfort, passive safety etc., see Chap. 7), but is the only way to optimize the product and keep development processes efficient at the same time.

1.3 The Product Evolution Process (PEP)

The PEP, also referred to as the *time-to-market process*, summarizes all activities for design and testing of the product as well as the set-up of production processes required for the manufacturing of the product. It is one of the three core automotive processes.[5]

In order to steer their vehicle projects, every OEM has their own detailed process model for the PEP, the "secret recipe" for their product development. These models define phases with milestones, specify the deliverables which are due at the respective milestones, and describe process chains as the participating players in the PEP and their roles. Although every corporate PEP is different (and usually strictly confidential), there are common patterns behind them representing an industry-wide accepted structure of vehicle development.

To discuss the PEP thoroughly, it must be regarded from different viewpoints (see Fig. 1.3): Seen from the timeline, we must distinguish the different phases of the PEP, from strategy to series support. From a process point of view, we must distinguish component design processes from integration processes and support processes. And applying the V-model of product development (see Sect. 1.3.3) we must always be clear whether we are in a phase of system design or integration.

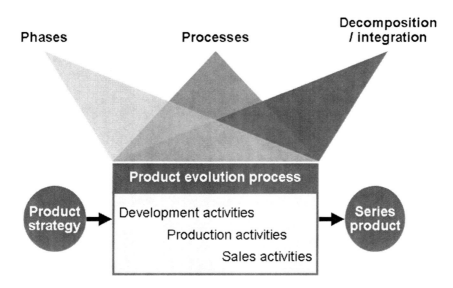

Fig. 1.3 Basic model of the PEP

---

[5] The other two core automotive processes are the *time-to-customer process* (that starts with the car being ordered by the customer and ends with the car being delivered to the customer) and the *time-for-service process* (that starts with the customer entering the dealership for service and ends with the customer leaving the dealership with his car in order).

1.3.1 Phases of the PEP

The first task in developing a new vehicle is the deployment of a *product strategy* or the general consideration of which cars a company should bring on the market at which point in time. Creating and updating this product strategy represents the continuous long-term planning process out of which distinctive vehicle projects or project programs are initiated. Figure 1.4 depicts how product strategy serves as a trigger for vehicle projects.

While product strategy is about complete vehicles, *pre-development* as another parallel continuous process deals with components and technologies (see Fig. 1.4). Here, innovative ideas taken from internal or external research, suppliers, partners or customers are concretized and evaluated regarding their technical and economic feasibility in products or related production processes. The decision which car will be the first to carry a pre-developed innovation is triggered by product strategy.

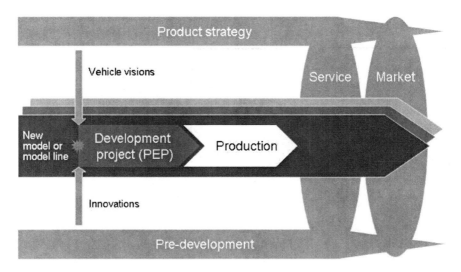

Fig. 1.4 Process framework of vehicle development projects

Both product strategy and pre-development are continuous processes and thus not project phases – even though they are often referred to as "strategy phase" or "predevelopment phase". Both processes are discussed in detail in Chap. 1.

Even if naming may vary among OEMs, the PEP generally is divided into three main phases: Initial phase, concept phase and series development phase. Figure 1.5 depicts this general vehicle project timeline together with the respective milestones and general objectives. In addition, Fig. 1.5 shows series support and further development as a fourth phase after *start of production* (SOP). The distinct phases of the PEP are discussed in detail over the course of Chap. 1.

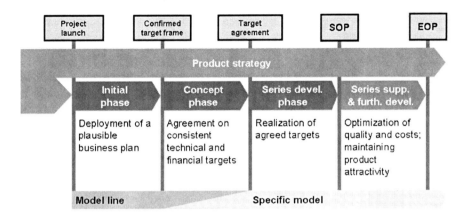

Fig. 1.5 Main phases of a vehicle development project

1.3.2 Processes of the PEP

1.3.2.1 Component Design Processes

Design *centers of competence* (CoCs) are home of the experts for specific parts and components. In a seat design CoC e.g., experts can be found for conceptual and series seat design, for testing, for cooperation with seat suppliers etc. By means of these *component design processes*, specifications are converted into series components by defining geometry, specifying material, planning a manufacturing process, and eventually releasing the component for series production and usage in a series product. With the release at the end of the process, the designer takes personal liability for the ability of the part to fulfill the specified requirements. On an upper level, most OEMs differentiate between the following six component design areas:

- *Powertrain design*: Engines, gearboxes, differentials, propulsion shafts, drive shafts, cooling system, exhaust system
- *Chassis design*: Axles, suspension, steering, pedals, wheels and tires
- *Body design*: Body structure (front, bottom, rear), outer panels, doors, hood, tailgate, fuel flap
- *Exterior trim design*: Front-end, bumpers, front/rear window, door system, trim parts etc.
- *Interior trim design*: Cockpit, trim parts, carpet, seats etc.

- *E/E component design*: Sensors, actuators, wiring and control units for driver assistance systems, information and communication systems, safety systems etc.

Organizationally, design areas are represented by design divisions, usually headed by a president directly reporting to the board of operations. Due to the geometric dependencies however, most OEMs have body and exterior trim design combined in one division. Other combinations are also possible.

Reporting to the presidents are their vice presidents, who manage the design CoCs. Within the chassis design division e.g., there might be one CoC for axles and wheels, one for brake systems and one for steering systems.

1.3.2.2 Complete Vehicle Integration Processes

Integration processes are used to define and develop the characteristics of the total vehicle. In contrast to the design engineers in the CoCs, integration engineers do not design any particular part, but rather evaluate the desired complete vehicle characteristics and feed their findings and technical recommendations back to the component processes. In doing this, integration processes steer the component design processes. The list below includes the six major integration processes.

- *Geometric integration* is the distribution and control of available space for all the vehicle's components. It embraces creation of the total vehicle package, allocation of package space to the component development CoCs, and monitoring of the geometric integrity of the complete vehicle by managing collisions and clearances (see Sect. 4.2).
- *Functional integration* is the validation of the functional characteristics of the complete vehicle from the customer's point of view, e.g. agility, cabin comfort, passive safety, etc. (see Chap. 7).
- *Systems integration* is the functional integration of the complete vehicle E/E system: Management of requirements, configuration and change management and integration of software with regard to development, production and service. Due to its criticality over the last decade, system integration is treated separately from functional integration (see Sect. 5.2.6).
- *Production integration* is the validation of the vehicle characteristics concerning production as well as the provision of the required production environment (see Sect. 8.1).
- *Service integration* is the validation of the vehicle characteristics concerning service, e.g. suitability for repair and maintenance (see Sect. 8.2).

1.3.2.3 Support Processes

To be able to develop products, design and integration processes need additional support processes. *Human Resources* has to provide people with the required skills at the right time and place in a development project. Especially the selection of the members of the project management team is crucial for the success of a development project. *Finance* has to check and provide budget and control project expenses to ensure cost stability. *Purchasing* selects capable suppliers for purchased parts or engineering services and contributes to the financial well-being of projects by analyzing and negotiating prices.

An important – though usually underestimated – task in a vehicle project is *internal communications*. Providing all members of the project team not only with the necessary information but also with internal news or success stories, internal communications can form a true team spirit and thus support successful project realization [3].

1.3.3 The V-Model of Product Development

Being the established process model in systems engineering (see Sect. 5.2), the V-model as shown in Fig. 1.6 allows a deeper understanding of the interplay of creative and analytical processes over the course of a vehicle development project. Starting with the specification of the desired complete vehicle characteristics, downward movement in the V-model (along the first leg of the V) denotes decomposition and specification - from complete vehicle requirements to system design and simulation down to parts specification, design and evaluation of parts. From here upwards (along the second leg of the V), the systems created out of the designed components are tested and validated against their specification in a hierarchical order - from components over sub-systems up to the complete vehicle. Design and validation of systems happens at the same level in the V-model [4].

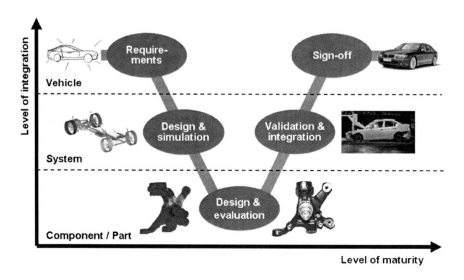

Fig. 1.6 Application of the V-model to the PEP (Source: BMW)

In contrast to IT-systems, automotive development requires several stages of prototype build and test at different points of product maturity. Each prototype build phase (see Sect. 3.3) represents a small product realization process itself. Hence, the V-model for automotive development shows subordinated Vs.

1.4 Vehicle Project Management

The task of Vehicle Project management is to organize and manage resources (such as money, people, materials, energy, space etc.) in such a way that the project is completed within defined targets (scope, quality, time and costs). It includes planning, controlling and deciding during product and process development. According to IEEE 1490 (2003), the nine disciplines of project management are [5]:

- Integration management
- Scope management
- Time management
- Cost management
- Quality management
- Human resources management
- Communications management
- Risk management
- Procurement management

As aforementioned, vehicle development projects are structured by milestones with defined deliverables attached. In contrast to other project management approaches, automotive project management usually sets milestones which require all processes engaged in the project to come to a consistent common state of product development.

A good example for a milestone and the respective project state is the start of a prototype build group. At the day set, the design CoCs must release a consistent set of parts that have been signed off by the integration processes. Also, production has to have their processes coordinated and matched to the released vehicle exactly on that day. Only this approach allows building prototype vehicles with the highest quality and hence the highest information value possible at this point in development.

With projects being structured by a series of these milestones, all activity is synchronized accordingly. For this reason, they are also called *synchro-points*. To ensure all sub-processes are on the right path between the synchro-points, they are structured and reviewed by *mini-synchro-points* (see Fig. 1.7).

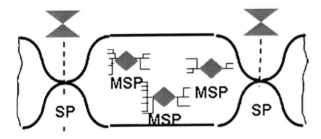

Fig. 1.7 Structuring of the PEP by synchro-points and mini-synchro-points (Source: BMW)

1.5 Aspects of International Development Projects

To remain competitive in an ever more global business environment, automotive companies increasingly work across geographical, socio-cultural and technical borders. Acting globally allows these companies to [6]:

- Increase turnover by selling and servicing their products
- Efficiently produce their products by utilizing local production technology and human resources
- Reduce transport costs and import duties by sourcing components locally
- Reduce risks from currency fluctuation by spending money in the same currencies as they earn it (natural hedging)
- Better design products to meet the local needs by involving local engineers

- Improve market position by expanding the range of potential suppliers and development partners

This strategic imperative for globalization in the automotive industry requires design engineers both at the *original equipment manufacturer* (OEM) and at the supplier sides to collaborate: to be able to sell cars in a new market, their compliance with local conditions and regulations must be considered and evaluated (a navigation system for instance, can only be tested in the country for which the car is developed and built). When launching production of a new car in a foreign plant, product process optimization requires local engineers to be involved in the problem solving process. And similarly, the local suppliers' engineers must be integrated in the vehicle development process executed at the OEM's engineering center. *Collaborative Engineering* is one of the keys to fully utilizing the business potentials of globalization. To make it work in reality, hindrances must be identified and dealt with.

The first of the two main hindrances are *socio-cultural differences*. Culture is apparent at first contact through language, clothes, habits, and then more pervasively through thoughts, unspoken assumptions, and values. In American-German teams e.g., the basic psychological differences are related to individualism versus collectivism, and to the avoidance of insecurity. Individualists focus on themselves, whereas collectivists put the group ahead of the individual. Americans are more individualists than the Germans. In terms of avoiding insecurity, Americans have much less need for security, and have more confidence in individual entrepreneurial thinking and personal abilities. Germans are formal, and want to use a systematic process, Americans are informal, and want to improvise. Germans want to take their time to reach a decision; Americans want an immediate result, even if based on a minimum amount of information.

These differences appear again in the language and communication process: In general, Americans are simpler, precise, informal, humorous, and friendly; Germans are more complicated, detailed, formal, reserved, and direct.

Finally, when one considers communication across the Atlantic, the distance creates barriers that have to be recognized and overcome. For instance, there are no chance meetings in the hallways to discuss a certain topic. The partners have to make the effort to establish the communication channels. Communication media does not lend itself to easily picking up the mood of the audience, and misunderstandings often occur because of the incomplete translation of thoughts into text using emails for instance. Thus, collaboration requires that the parties involved increase their communication efforts and are aware of the possible problems, and therefore take a proactive role in clarifying points, and asking for detailed explanations to avoid misunderstandings.

Aside from the "soft" socio-cultural differences, there are *differences in the business environment*: Technical, educational and legal facts that must be considered to ensure successful international cooperation:

- Technical border conditions: Climatic differences may negatively influence the transfer of a manufacturing process such as gluing; Electric power supply differs not only in voltage but also in frequency and stability; Professional education of workers might require more or less detailed description of how to do it in different countries.
- Materials and standard parts: Engineers in different countries prefer different materials – and have collected specific know-how concerning making, processing and designing with these materials. Although other materials might be also available, they usually are expensive and considered somewhat exotic. Standard parts such as fasteners, actuators, hoses, sealants differ in terms of size, geometry and specification/performance.
- Development standards: Material properties, tests, tolerances and manufacturing processes are specified by different standards, e.g. in Europe according to DIN/EN/ISO and in the U.S. according to SAE, ASTM.
- Legal background: An important boundary condition for international collaboration is the legal system providing the applicable laws. To ignore the peculiarities e.g. of the U.S. legal system can be extremely high risk for foreign companies, not only concerning product liability. Other fields of important legal discrepancies are patent law, health care or business taxation.

In summary, companies have to consider the various differences between the cultures, the technical and legal aspects of doing business, and must weigh the gains that may result from the collaboration against the overcoming of the hindrances to collaboration outlined above. On the other hand, collaboration brings forth the collective thinking process of people with different backgrounds and experiences. The richness of solutions is typically expected, and the tailoring of the products for the markets in which they are developed enhances market acceptance and penetration and therefore company reputation and market share.

References

1. Holzbaur U (2007) Entwicklungsmanagement. Springer, Berlin
2. Cusumano M, Nobeoka K (1998) Thinking beyond lean. The Free Press, New York
3. Hausen-Mabilon F (1999) Konzeption einer Strategie zur Verbesserung der internen Kommunikation am Beispiel einer Projektgruppe in der Automobilindustrie. Dissertation.de, Berlin
4. Rausch A, Broy M (2007) Das V-Modell XT. dpunkt, Heidelberg
5. IEEE 1490 (2003) Adoption of PMI standard. A guide to the project management body of knowledge – description
6. Weber J, Fadel G (2006) Opportunities and Hindrances to collaborative automotive development. SAE Transactions 2007, 115(5):911–920

Chapter 2
Product Strategy

Abstract The basic and most important decisions of any automotive OEM relate to the question of when to bring which vehicle to the market. Even though there is no formula for guaranteed success, an analysis of cars that succeeded and cars that flopped in their respective markets leads to a list of minimum requirements that should be checked as part of the of strategic decision making process.

2.1 Cars that Topped and Cars that Flopped

The task of automotive product strategy is nothing less than to give the best possible prediction of which cars the customers will buy in the future. The best development processes can not compensate for wrong strategic assumptions. This makes product strategy the most important task in vehicle development and the driving force in corporate strategy. The answers product strategy has to give include:

- Which cars will customers buy in the future – and how many?
- What should the complete product portfolio look like in terms of brands, model lines, variants?
- Is it enough to continue and redesign or are new models or model lines required?

Viable predictions to these questions require comprehensive consideration of future boundary conditions such as:

- Customer needs: Will certain features such as high speed and dynamic performance still be important, when megacities grow and urban and suburban streets get ever more congested?
- Social acceptance: What will society think about cars with any kind of emissions? Will the whole principle of individual mobility still be generally accepted?
- Brand values: Will the current brand values still appeal to future customers?
- Legislation: Which laws and regulations will apply worldwide that might influence the purchasing decision or development, manufacturing and sales processes?

J. Weber, *Automotive Development Processes*, DOI 10.1007/978-3-642-01253-2_2,
© Springer-Verlag Berlin Heidelberg 2009

- Corporate strategy: What is the long-term plan for the company regarding location, employees, technologies etc.?
- Competition: Which cars will the competition offer?

Automotive history is full of examples for cars which were eminently successful at their time or on the contrary just flopped. With the wisdom of hindsight, it is usually very easy to analyze the respective reasons. Analysis of these projects is important to be able to deduce the interrelations that lead to failure or success.

2.1.1 Tops

There are many definitions of what success means for automotive development: Number of styling awards, ranking in quality assessments or customer surveys, ROI etc. In terms of strategy however, the relevant question is not so much the detailed result of series development, but the coherence and consequent long-lasting attractiveness of the general vehicle concept. From the viewpoint of an OEM, a commonly accepted measure for the success of a vehicle concept is hence the number of units based on that general concept that could be sold over time. With this in mind, the five most successful cars in automotive history are the Toyota Corolla, the Ford F-Series, the Volkswagen Golf, the Volkswagen Beetle and the Ford Model T [1].

The first car that was manufactured on an assembly line, the Ford Model T started a new era of the automotive industry. 16,500,000 cars were sold from 1908 through 1927. The commercial success however stemmed from the new production approach that allowed an unrivaled price. In 1914, assembly time was only 93 minutes and the Model T was sold for $850 (and later even below $300) when competing cars were priced at over $2,000. The lead over competition was so big, that no advertising was needed for the Model T between 1917 and 1923.

At the end, the reason for its success became the reason for its end: Sticking to the same concept to allow fast and efficient production. By 1925, other cars offered much more comfort and style – now at competitive prices. The Model T lost its supremacy on the market and Ford discontinued production in 1927.

Another car that never really changed its initial concept is the legendary Volkswagen Beetle (see Fig. 2.1 top left). 21,529,464 units produced between 1935 and 1983 make it the fourth best selling car in history. Compared to its competitors such as Citroen 2CV, the Beetle had superior performance,[6] excellent handling and was still a low cost car both for purchase and for maintenance. Together with the unique body style, this made the Beetle a trendy and desirable car for generations of customers.

---

[6] Max. speed 115 km/h (72 mph), acceleration 0–100 km/h (0–60 mph) 27.5 seconds; fuel consumption 7.6 l/100 km (31 mpg) with a standard 25 kW (34 hp) engine.

Fig. 2.1 Most successful cars ever: VW Beetle, VW Golf, Ford F-Series, Toyota Corolla
(Sources: Volkswagen, Ford, Toyota)

In 1974, when popularity of the Beetle started to decline, Volkswagen launched their second big hit: The VW Golf (see Fig. 2.1 top right), which gained world-wide popularity through five redesign generations. Although the concept never was really accepted by the big American market, to date more than 24 million Golfs have been sold world-wide.

The success of the Golf is founded in providing a new concept in the compact class (water-cooled front-wheel drive, east-west engine, hatch-back) that was affordable for everyone, offered enough space for five passengers and baggage and still was sporty and considered cool. The Golf was so dominating in the compact class that it is still called the *Golf Class*. In Germany, people born between 1965 and 1975 are referred to as the *Generation Golf*.

While the golf never was a big selling car on the U.S. market, the Ford F-Series Pick-up truck (see Fig. 2.1 bottom left) has sold over 25 million units since 1984, solely in North America. The F-Series has been manufactured for over 5 decades and is now in the 11th generation, and has been the best selling vehicle in the U.S. for 23 years. The basis for this success is its high reliability gained by use of robust technical solutions, the huge choice of body and trim options and the affordable price.

Toyota introduced the Corolla (see Fig. 2.1 bottom right) in 1966. While styling initially was rather unexciting, the Corolla convinced its customers by quality and cost-effectiveness. Toyota has kept the car attractive now for over 40 years and completely redesigned each of the Corolla's nine generations. It was first in its class in almost every quality and reliability ranking. To date, over 35 million Corollas have been sold all over the world.

2.1.2 Flops

The reasons for the success of the top-selling cars listed above are manifold – and so are the reasons that lead to a vehicle not coming close to reaching the planned sales targets, commonly called a market flop. Again: Market success is much more than design quality! Some of these flops are or were technically brilliant car concepts that were just realized too late or too early, had wrong assumptions of future customer priorities set as a basis for design, relied too much on the attractiveness of innovations, or were even just poorly marketed. Four prominent examples of vehicles that were loved by the people who bought them – but from their manufacturers' point of view did not convince enough people to buy them – are the Ford Edsel, the Renault Avantime, the Glas 2600 and the GMC Envoy XUV.

The probably best-known and most spectacular flop in automotive history is the *Ford Edsel* (see Fig. 2.2 top left), which was manufactured by Ford Motor Company from 1958 to 1960. It is a good example, because it was not poor quality or concept but a series of circumstances that eventually led to failure. It was "the wrong car at the wrong time with an awkward name and was too big when economical circumstances demanded for smaller cars" [2].

An example of a car that was flawless and well-accepted by customers but still became an economic disaster is the *Glas 2600 V8* (see Fig. 2.2 bottom left). In 1966, the product strategy of Glas, well-known for designing the after-war Goggomobil in the 1950s, crossed the boundaries of its brand by designing the Glas 2600 V8, a technically brilliant sports car which was nicknamed "Glaserati" at its time. But the overall structure of the Glas company was not ready for cars at this high level. Production expenses grew too high and only a pre-series was built of the 2600 V8 when the financial situation of Glas was so bad that they were bought by BMW in 1966.

A more recent example is the *GMC Envoy XUV* (see Fig. 2.2 bottom right): Despite a highly innovative retractable roof, the XUV sold so poorly that production was discontinued after only 2 years. It was never really obvious, why customers more or less ignored this car.

The *Renault Avantime* (see Fig. 2.2 top right) was designed and built by Matra between 2001 and 2003. It had a radical and unique design, a mixture of van and coupe. Though other ambitious Renault designs at the beginning of the 21st century were very successful, customers did not appreciate the Avantime at all. Sales were extremely poor and the project became somewhat of an economic disaster. As a result, Matra went bankrupt and pulled out of the automotive production business in 2003. Renault decided to discontinue the Avantime – after only 8,545 cars were built.

Fig. 2.2 Cars that flopped: Ford Edsel, Renault Avantime, Glas 2600, GMC Envoy XUV
(Sources: Ford, BMW, Renault, Wikipedia)

2.2 Factors of Success in the Automotive Industry

Detailed reflection on the tops and flops mentioned above fosters the understanding of the prerequisites of successful vehicle projects. Product strategy determines the four essential drivers for a vehicle's market success [3]: Worldwide market presence, a consistent model mix, a sharp brand profile and a product profile that suits the needs of its potential customers.

2.2.1 Worldwide Market Presence

How could Toyota sell over 35 million Corollas? Apart from being a practical, reliable and affordable car, it suited from its very beginning the needs of almost every market. This is achieved by Toyota's widespread worldwide production network. 53 overseas (outside Japan) manufacturing companies in 27 countries / regions and development centers in North America, Europe and Asia [4] allow customization of base vehicles to local requirements and preferences.

While customers apparently expect cars to be tuned to local preferences in the mass market, they obviously demand something like a world-wide accepted level of design and styling in the premium sector. A BMW X5 e.g. is – apart from possible differences due to local registration legislation – sold in exactly the same way all over the world. And it is manufactured as a world-model solely in one plant.

Equally, the new MINI is a world model, exclusively produced in Oxford, England. When it was launched in 2001, a tremendous marketing effort successfully positioned it as an international car that even created an international culture and society of MINI drivers.

Today, the global automotive market can be split in two main parts: Emerging markets such as China or India show impressive growth rates while vehicle sales in the triad (North America, Europe and Japan) seem to have reached their limit. In absolute figures on the other hand, the triad is still expected to represent about 70% of world-wide vehicle sales in 2011 [5]. Figure 2.3 shows sales of cars and light trucks from 2003 to 2007 and a forecast until 2011 for selected markets.

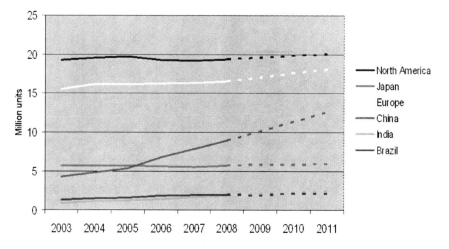

Fig. 2.3 Vehicle sales in selected markets (Data source: Automotive News [5])

2.2.2 Model Mix

Just like balance and consistency are the basic criteria for investment portfolios in order to achieve sustainable financial success, they are also prerequisite criteria for the product portfolio of an automotive manufacturer in order to allow long-term market success. The collection of models a manufacturer offers should be well-balanced between market requirements on the one hand and development and production requirements on the other: From a market point of view, there must be a complete and consistent model mix, clearly differentiated by functionality, size, performance or level of comfort while at the same time maintaining a distinct common brand profile. If the portfolio expands over more than one brand – which is usually required to keep brand profiles sharp when extending the product portfolio – brands must be clearly differentiated by their brand values. Additionally, from a commercial point of view all models should share as many parts and components as possible in order to optimize development effort, purchasing volume and production flexibility.

As an example, the major reason for the serious crisis BMW went through in 1959 and which almost led to the company being sold to Daimler Benz, was the lack of medium sized cars in its product portfolio which had been spread out by the minimalist Isetta on the one end and a series of luxury V8 cars on the other end (see Fig. 2.4). The brand BMW was rather blurred and ambiguous instead of creating a clear picture of brand values in the customers mind.[7] In addition, the obvious technical differences of both cars made common usage of parts or equipment impossible and led to losses of about 4,000 Deutsche Mark for every 507 sold.

Luxury

Upper middle

Middle

Compact

Small

Micro

Fig. 2.4 BMW product portfolio in 1959 (Source: BMW)

---

[7] German magazine "Der Spiegel" wrote in 1959 that BMW would build cars for bank directors and day-talers.

After filling the gap by successfully introducing the 700 in 1960 and subsequently the 1500, 1800 and 2000, BMW recovered and strategically develop their product portfolio since. Figure 2.5 shows the current vehicle portfolio of the BMW Group.

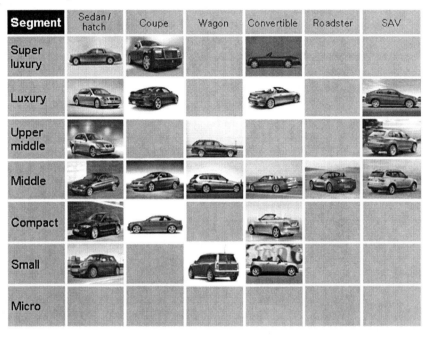

M-Models:
M3 Coupe, M3 Convertible, M5 Sedan, M6 Coupe, M6 Convertible, Z4 M Roadster, Z4 M Coupe

Fig. 2.5 BMW product portfolio 2007 (Source: BMW)

While leaving gaps in the portfolio can seriously harm the brand profile, an overstuffed and overlapping model mix blurs differentiation. With the Golf Variant and Bora Variant, Volkswagen in the early 2000s offered two wagon type cars of the same brand in the same segment. As customers did not realize a significant difference between the two cars, practically no additional volume was sold – and the additional effort to develop, build and sell a second car was wasted. In the same way, lack of demarcation between the brands of one company also diminishes the sales effects that are intended by offering different brands. Here also, Volkswagen is an example. Within their platform strategy, differentiation between their brands was partly blurred, leading to in-house competition and volume cannibalization: Customers were aware that e.g. a VW Golf and a Skoda Octavia shared the same platform, which made some customers opt for a fully loaded Octavia rather than VW Golf without any options for the same price.

2.2.3 Brand Profile

The brand profile is the set of values and emotions that a potential customer associates with a brand. It is created by public perception of the products, their marketing and the company itself over many generations. In the automotive business, the brand image can be even more decisive for a customer's decision than technical facts. For most customers, it is an important asset of their car which they are willing to pay for. Figure 2.6 indicates the automotive brands among the top-20 most valuable brands world-wide.

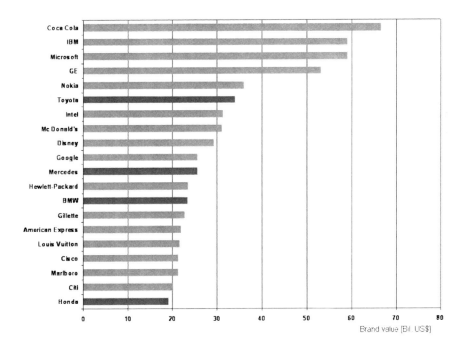

Fig. 2.6 Most valuable global brands 2008 (Data source: Interbrand [6])

Hence, the first criterion for product planning is conformity with the brand profile, which is created and maintained by marketing in a long-term process to be well-profiled and desirable. If the publicly perceived brand profile diverges from the intended one, measures can only be long-term.

A good example for a brand profile being a set framework for new products is Mercedes-Benz's expansion of their product portfolio to the middle class segment by introducing the 190. Until then, their product portfolio only covered the upper and upper middle class segment. Management was reluctant to introduce a middle class car because it feared this could dilute the strong brand and adversely affect appreciation and sales of existing models. When eventually the 190 was launched in 1982, it was a huge success and actually did not harm sales in the upper segments of Mercedes Benz in any way. Mercedes Benz expanded their portfolio even further downwards to the compact class with the introduction of the A-Class in 1997. Only the micro car that was brought to the market in 1998 was apparently incompatible with the Mercedes-Benz brand and consequently was introduced as the Smart.

An example where – in the perception of potential customers – the capacity of a brand has been stretched too far is the VW Phaeton, introduced in 2002. Though a technically excellent car, upper segment customers never really could bring themselves to buy a Volkswagen, a brand that is a synonym for affordable mass market "people's cars". Sales were well below plan, especially in North America where the Phaeton was discontinued in 2006.

2.2.4 Product Profile

While a brand profile describes the long-term perception of a brand, the product profile specifically describes targets for a new car or model line. The product profile has three main aspects: Product position, target group and product properties. All three must follow technical, legal and societal trends. Product profiling is often supported by technical and economical analysis of competing vehicles. Figure 2.7 shows the new MINI surrounded by vehicles that represent its target market at the time of its development.

Fig. 2.7 Competition analysis for the New MINI (Source: BMW)

Product positioning must make it unambiguously clear, in which segment the new car competes and in which categories the car must be best in segment or at least top three to gain its intended market share, e.g. by being the fastest, most economic, cheapest, youngest, coolest, smallest etc. car in segment.

When describing *target groups*, global and local trends such as demographic shifts, changes in purchasing power, and changes in consumption behavior have to be considered. The demographic prognosis shown in Fig. 2.8 e.g. indicates that from 2020 on, older customers will play a significantly greater role, which makes them and their specific requirements ever more important for strategic product placement.

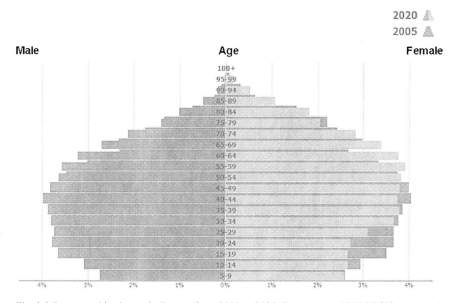

Fig. 2.8 Demographic change in Europe from 2005 to 2020 (Data source: UNECE [7])

Another so called megatrend is the shift in purchasing power from Europe to Asia. Forecasts show that between now and 2050 buying power will dramatically increase in Asia, will stay about the same in North America and decrease in Europe. Other trends that definitely influence customers are the ongoing move into ever larger mega-cities, or the increasing interest in environmental friendly cars.

Last but not least, the profile of the future car must also indicate its physical *product properties*. Here, the first step is to determine the conceptual architecture in terms of body concept, proportion, and exterior and interior style. Compliance with future legal requirements such as emission / fuel consumption limits or safety standards has to be verified and market specific conditions like climate, traffic conditions or even driving habits have to be reflected in the product profile.

The next step in profiling the new car is determination of characteristics (such as vehicle dynamics or level of comfort and safety) as well as a catalogue of basic and optional equipment, bundles and special editions. This also includes the product innovations, of which the most differentiating are marked as *unique selling propositions* (USPs) for marketing of the car. The profile is rounded by planned values for price and cost of ownership.

References

1. Automotoportal (2006) Top 5 world's most successful cars ever. http://www.automotoportal.com/article/Top_5_Worlds_Most_Successful_Cars_Ever. Accessed 20 November 2008
2. Deutsch J (1976) Selling the people's Cadillac: The Edsel and corporate responsibility. Yale University Press
3. Theissen M (2007) Innovative Produktentwicklung. Lecture hand-out, Hochschule für Technik und Wirtschaft Dresden
4. Toyota (2008) Company profile. Manufacturing. Worldwide operations. http://www.toyota.co.jp/en/about_toyota/manufacturing/worldwide.html. Accessed 20 November 2008
5. Automotive News (2007) 2007 Global Market Data Book. Crain, Detroit (MI)
6. Interbrand (2008) Best Global Brands. 2008 Rankings. http://www.interbrand.com/best_global_brands.aspx?langid=1000 Accessed 20 November 2008
7. United Nations Economic Commission (2006) World population prospects: The 2006 revision population database. http://esa.un.org/unpp. Accessed 20 November 2008

Chapter 3
Phases of the Product Evolution Process

Abstract After product strategy has created a vision of what kind of vehicle should be developed, a formal development project is started. The PEP – as the master process model for such projects is commonly structured in three main phases: During the initial phase, the technical and economical feasibility of the project is clarified, which is concretized over the concept phase to a consistent set of targets. The task of the series development phase is then to realize the vehicle so that it can be produced by a manufacturing plant. Not a distinct phase of the PEP but also a model related design task is post-launch series support and further development.

3.1 Initial Phase

During the initial phase, the product profiles coming out of product strategy are concretized towards a consistent target framework from which the technical and economic feasibility of a vehicle project as a whole can be evaluated (see Fig. 3.1). The target framework must be plausible, prioritized and its conflicts resolved.

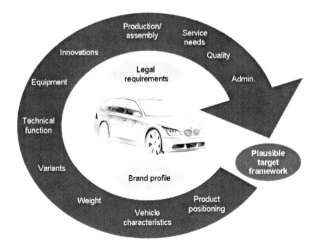

Fig. 3.1 Target framework for a new model (Source: BMW [1])

J. Weber, *Automotive Development Processes*, DOI 10.1007/978-3-642-01253-2_3,
© Springer-Verlag Berlin Heidelberg 2009

3.1.1 Technical Feasibility

Major aspects of the technical feasibility of a vehicle project include the geometric verification of the vehicle package, initial – exterior and interior – styling decisions, and early considerations concerning production.

The vehicle design process – which is discussed in detail in Sect. 7.3 – starts during the initial phase. The general body concept is selected, the major components are derived from the planned vehicle functionalities and are roughly positioned in the vehicle package, and first drafts of exterior and interior styling are created. Part of this initial packaging process – which is usually done in 2D – is the analysis of critical dimensional chains and the resolution of their target conflicts. As an example, Fig. 3.2 shows the dimensional chain that starts at the top of the rear roof panel and then goes through roof liner, the rear passengers' head, torso and buttocks, rear seat foam, floor panel, sound insulation, fuel tank and under floor protector. Along this dimensional chain, a variety of requirements leads to target conflicts that have to be resolved:

- For dynamic vehicle appearance, styling requires a low roof contour.
- Reduction of interior noise requires a thick roof liner and thick sound insulation attached to the lower body panels under the rear seat.
- Passenger ergonomics require that an average adult sitting in the rear has enough headspace.
- Ride comfort for rear passengers is mainly determined by the thickness of the seat cushion foam.
- As the fuel tank is typically located between the lower body sound insulation and the under floor protector (which demarks the vehicle's bottom contour), their distance determines the fuel tank volume and thus the cruising range of the vehicle.
- The ground clearance (that is the distance between road line and the lowest vehicle part, determines complete vehicle characteristics such as agility, aerodynamic resistance and uplift, or off-road-capability.

Apart from the directly vehicle related characteristics, the target framework also must give answers to questions regarding planned production, procurement and service processes: One or more production plants have to be selected (or expanded or even built) that match the project's requirements in terms of production location, technological capability, volume capacity and supplier network. Likewise, the new vehicle must either fit into existing sales and service structures or these structures have to be expanded or newly created. Required marketing measures have to be pre-estimated. For the first model of a new or reanimated vehicle brand, a complete marketing concept has to be planned.

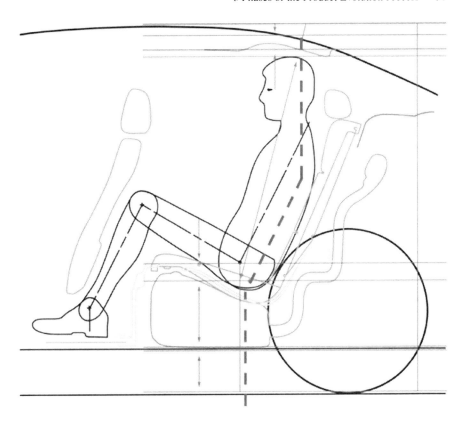

Fig. 3.2 Critical dimensional chain through the rear passenger compartment (Source: BMW)

The product target framework also includes a commonality structure that determines which parts or components are reused from existing models (*new design* or *carry over part*).

If the development project includes a model line, the target frame must include – and must be consistent for – all variants of that model line. Its commonality structure must fix which parts are used exclusively for base vehicle or only one variant (*solitaire parts*) and which are used for more variants (*communal parts*). Hence, within a model line, the initial phase for all variants starts concurrently with the base model, independent from when the following phases for the variants will actually start.

3.1.2 Economic Feasibility

At the end of the initial phase, a consistent and feasible business plan for the new model or model line must be available, based on plausible assumptions concerning all types of earnings and costs that occur over the expanded lifecycle of a vehicle (see Fig. 3.3). Only if – based on these figures – a certain *return on investment* (ROI) or *net present value* (NPV) can realistically be achieved, will the project be continued and led over into the concept phase. The business plan is usually worked out by a team with representatives from corporate strategy, marketing, development, purchasing, production and finance.

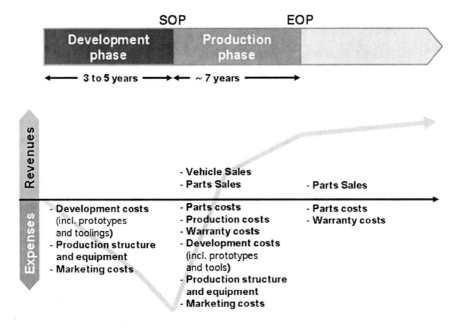

Fig. 3.3 Expenses and revenues within the expanded vehicle life cycle

The sample values listed in Table 3.1 should help in creating a realistic impression of the financial dimensions of a simple vehicle project:

Table 3.1 Sample financial values for a vehicle development project

Sales price per unit	43,000 €
Planned total volume	250,000 units
Period of production	84 months / 7 years
Investment	700 Mio €
Development costs	400 Mio €
Manufacturing costs per unit	16,000 €

With these example values, the NPV as the main financial indicator of the business case would calculate as 3,000 Mio €. Figure 3.4 shows expenses and revenues for each year and the corresponding NPV curve.

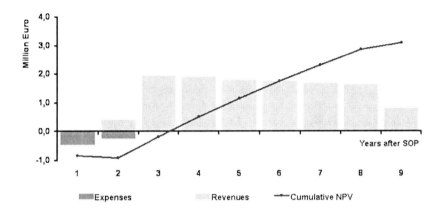

Fig. 3.4 Expenses, revenues and NPV progression after SOP for a sample vehicle project

As the resulting NPV is one of the most important indicators on which the decision whether to proceed with a project or not is based, it is important to know not only its (forecast) value but also its sensitivities. The following deviations from the assumed values according to Table 3.1 would all lead to an increase or decrease of the NPV by ± 500 Mio. €:

- Sales price ± 500 €/unit (± 1.1%)
- Sales volume ± 10,000 units (± 4.0%)
- Manufacturing costs ± 500 €/unit (± 3.1%)
- Investment ± 70 Mio. € (± 10.0%)

3.2 Concept Phase

3.2.1 Vehicle Concept Design

As during the concept phase the vehicle must be detailed far enough that it is clear whether it will – when realized – meet the set requirements, virtual and real 3D models are the main form of product representation during the concept phase (see Chap. 4 for the full *virtual car process*). Exterior and interior surfaces are specified first in clay models and only then transferred into computer aided design (CAD) data (see Sect. 7.3 for the full *vehicle design process*). 3D CAD models are created by design CoCs and analyzed by the integration processes (see Sect. 1.3.2.2) in order to evaluate complete vehicle characteristics and thus validate the vehicle concept. Figure 3.5 shows examples for the virtual evaluation of driver ergonomics, aerodynamics, handling, and production. These and other complete vehicle characteristics are extensively discussed in Chaps. 7 and 8.

Fig. 3.5 Vehicle concept validation based on virtual 3D models
(Sources: Human Solutions, BMW)

Even if most complete vehicle characteristics can be simulated on the basis of 3D CAD data, the conceptual validation, especially of vehicle agility, is performed by testing real experimental vehicles – current models in which chassis and parts of the powertrain have been updated to the concept design level of the new model.

While vehicle projects are managed by relatively small *initial teams* during the initial phase, responsibility is handed over to a comprehensive project organization at the beginning of the concept phase. The project organization commissions, monitors and steers decentralized concept teams and establishes the required meeting and reporting structures to enable optimum cooperation. The actual organizational models different OEMs use for structuring their model lines and vehicle projects are similar; they mostly differ in the strength of the project manager relative to the CoC managers.

3.2.2 Target Agreement

Based on the target framework and the business plan worked out during the initial phase, a detailed and coherent project concept with quantified technical and financial targets is delivered at the end of the subsequent concept phase. This project concept includes the following main sub-concepts:

- Product concept
- Production concept
- Sourcing concept
- Sales and marketing concept
- Service concept

To ensure quantitative consistency of the concept, target management methods and systems are applied (see Sect. 6.1). The consistent set of quantitative targets the project concept comprises is the basis for the important project milestone *target agreement* which marks the end of the concept phase and – as it triggers series development with all associated costs and risks – represents something like a "point of no return" in the vehicle project. At this point, management has to answer key questions such as:

- Is the car technically and economically feasible? Can all required functions be realized at the planned cost? Will the planned cars meet all internal and external/legal requirements?
- Are the targets for sales price and sales volume realistic? Are these the cars enough customers will be ready to pay for? Especially the ones that will be launched last?
- Is the supplier structure feasible? Will all potential suppliers have the experience and resources necessary to deliver on time in the right quality? Are all suppliers financially healthy enough to be able to deliver even in times of a crisis?
- Are production targets feasible? Can the new cars be built in existing production structures or can required changes be implemented on time and at planned cost without unplanned impact on currently running series production? Can enough people with the required skills be at hand when production starts?
- Are the project targets feasible? Can series development be carried through with the planned resources at the planned cost and in planned time?

Only if the answer is yes to all of these questions will management take the entrepreneurial risk and start series development of the first car of the model line – which means closing on supplier contracts and irrevocably releasing investments, e.g. for series tooling.

3.3 Series Development Phase

3.3.1 Component Design

Working out the product concept leads to detailed design that ends in formal releases, the unambiguous descriptions of parts (in terms of geometry, material, surface) and software (see Sect. 5.2.5). Releases are the binding base for suppliers
and trigger the manufacturing process for tools. The huge economical importance
of releases stems both from being the basis for individual responsibility of the releasing designer in terms of product liability and the investment costs necessary
for tooling.[8]

To ensure delivery of parts at planned cost, quality and reliability, suppliers are
selected and qualified at the beginning of the series development phase. Together
with the OEM, they design and realize the tools, equipment and processes necessary to manufacture the parts under series conditions.

3.3.2 Complete Vehicle Integration

The central mission during series development is the integration of components
and systems to a vehicle that meets the initially defined legal and self-imposed requirements regarding the complete vehicle characteristics. For this evaluation –
and eventual validation – prototypes, pre-series cars and series cars are tested [2].
Figure 3.6 shows typical tests for complete vehicle validation.

3.3.3 Prototype Build

In a complete redesign project, OEMs today typically build and test two prototype
groups: An early group with which the concept is evaluated and from which necessary design changes are derived, and a second one that incorporates these
changes and proves them successful. As long-running series tools for bigger parts
are released before the second prototype group is built, design changes stemming
from the test results of the second build group would be very costly and risk post-
ponement of the planned SOP.

The points in time at which prototype groups are built have to be selected carefully. On one hand, a later build allows having all the latest solutions incorporated
into the prototype while on the other hand an early build allows early test results

---

[8] Even if parts are manufactured by a supplier at his own site, the related tools are paid for by
and remain property of the OEM.

and leaves more time for possible design changes. Within a prototype group, the order in which the cars are built is determined by their users and the time-criticality of their tests. As e.g. a failed crash test can lead to comprehensive changes throughout the whole vehicle structure, the first vehicles in each build group are usually used in crash tests.

To obtain realistic test results, prototype testing can not be restricted to confined testing facilities. However, exposing body styling and other design features too early on public roads would not only allow competition to get an advantage by adapting their products accordingly but – even more important – would also make the current model less attractive in the perception of the customers and thus negatively impact sales volume and price. For these two reasons, prototypes are usually camouflaged; heavily in the beginning and only lightly at the end of development.

Fig. 3.6 Complete vehicle validation (Source: BMW)

Provision of prototypes for the evaluation of complete vehicle characteristics is one of the major cost factors in vehicle development. An early prototype can cost up to one million Euros. There are three main ways to reduce the number of prototypes required and their associated costs:

- Configuration. Through intensive negotiation among the different departments that have to test the prototypes, the combinations of body style, engine, gearbox, left-hand/right-hand drive, country version etc. can be found that require a minimum number of prototypes. For crash tests e.g., the configuration with the heaviest engine and the least rigid body is the most critical. Hence, the passive safety department will insist on a convertible with the biggest diesel engine.

- Multiple use. Prototypes used to be an informal status symbol for design departments, and some managers made sure that their CoC owned a prototype of every model. With companies now looking rigidly at their development costs, test drivers from all departments have to share a much smaller number of prototypes according to a precise time table.
- Simulation: Increasing accuracy and reliability of simulation tools can reduce the need for hardware prototypes.

While testing of prototypes delivers the required data for product development, building the prototypes is the first opportunity for production to evaluate planned production processes and the buildability of the car. Even if parts usually come out of prototype tools and assembly happens in a lab situation rather than in a real plant, prototype build uses and thus evaluates new equipment and at the same time evaluates the production processes for every single part according to the planned sequence. Figure 3.7 depicts sample processes of prototype build: Manual body welding in temporary fixtures (top left), measuring prototype components before assembly (top right), assembly of pre-assembled powertrain and suspension using close-to-production equipment (bottom left), and camouflaging of prototypes (bottom right).

Fig. 3.7 Prototype build (Source: BMW)

In order to prepare series production, the development activities are gradually transferred from the development center to the production plant (or plants) at the end of the series development phase. There, buildings, equipment, processes and trained resources for vehicle manufacturing have to be prepared in time for pre-series production, the first cars of a new model which are built in the plant.

3.3.4 Launch Preparation

The last part of the series development phase is the *launch phase* or *ramp-up phase*, which starts with the first pre-series car being assembled in the plant and ends with daily series production volume having reached its capacity target. SOP denotes the point in time after pre-series production that the first car that is intended to be sold to a real customer is produced.

On the sales side, an exactly planned roadmap is followed to launch marketing activities such as publishing teaser pictures, press presentation, staging of the vehicle's debut at car exhibitions or advertising campaigns. The first cars built during ramp-up production following SOP are delivered to dealers' showrooms as part of the marketing campaign. Parallel to that, sales and service staff are trained, showrooms and repair shops prepared and transport of the new models to their points of sale is organized.

An important milestone within the launch phase is the *ordering release*, which gives the dealers the go-ahead to file customer orders. At this point management must commit that all ordered cars can be delivered as specified, on time and in 100% quality. At the end of the *series development phase*, the following results should be achieved [3]:

- Product: All vehicle functions, features and properties defined in the target agreement are realized in the customer cars.
- Sales: Market introduction activities for vehicles and aftersales are finished and confirmed.
- Production: After production volume and quality targets have been reached, the production process capability is confirmed.
- Finance: The agreed economical targets are all met.

3.4 Series Support and Further Development

When series production has been proven stable (which means having run at planned capacity for a couple of weeks without major problems), the series development phase formally ends. The project team is discharged and responsibility is handed over to a much smaller series further development team.

But as stated in Sect. 1.3.1, development activities continue even after that. *Series support* comprises short-term measures to optimize costs and quality, component discontinuations and to fulfill short-term legal and insurance requirements. *Further development* includes all changes that can be planned on a longer term. A facelift e.g. is carried through once in the life of a vehicle and mostly includes design measures and new features or options in order to refresh the car on the market.

Measures that optimize quality, costs or weight or fulfill legal requirements are usually implemented during periods of planned production shutdown so that changed parts and changed equipment can be provided without interfering with production. Thus, cars that are built after these shutdown measures may be significantly better than the ones before and are differentiated by the model year.

References

1. Theissen M (2007) Innovative Produktentwicklung. Lecture hand-out, Hochschule für Technik und Wirtschaft Dresden
2. Sörensen D (2006) The automotive development process. Deutscher Universitäts-Verlag, Wiesbaden
3. Weber J (1999) Optimierung des Serienanlaufs in der Automobilproduktion. VDI-Z 141(11/12):23–25

Chapter 4
Virtual Car Process

Abstract The Virtual Car Process comprises all activities required to structure, build and test computer-internal representations of motorized vehicles – so-called virtual cars. Containing both geometric and functional data, virtual cars are not only the basis to facilitate decentralized design in context but are also the most important lever to increase efficiency in vehicle development.

4.1 Building Virtual Cars

4.1.1 Purpose and Benefits

Virtualization of products and processes is one of the areas in which automotive development has seen the most dramatic improvements over the last two decades. First 3D modelers helped designers to generate parts geometries – which then usually were converted back to 2D drawings. Over time, by attaching further geometric information (such as a part's position in a common work space, surface properties etc.), administrative data (such as part numbers, bill of materials (BOM) structures, versions, variants and options etc.) and especially functional properties (such as material information, strength, color etc.), 3D parts became virtual cars which became the central communication platform for the co-operative vehicle development process.

A virtual car is the representation of all of a specified vehicle's parts in a shared workspace that allows the simultaneous and coordinated development of the complete vehicle. In addition to being a platform for design-in-context, virtual cars represent the basis for simulation and evaluation of vehicle properties, functions, costs and weight. Usage of virtual cars enables fast prototype build and fast ramp-up by increased geometric consistency of parts and pre-checked assembly processes. Costly changes of hardware parts and manufacturing equipment can be dramatically reduced. Virtual cars substitute for hardware prototypes to an ever growing extent. In passive safety e.g., real crash tests – while still necessary – are expected to 100% confirm the findings from simulation (see Sect. 7.7.3).

J. Weber, *Automotive Development Processes*, DOI 10.1007/978-3-642-01253-2_4,
© Springer-Verlag Berlin Heidelberg 2009

The usage of virtual cars does not make personal communication unnecessary; rather it intensifies the need for it. The chassis designer who has worked several weeks to optimize his rear axle carrier and now wants to update it in the common workspace might realize that the body panel he used to take as a reference has just been changed and now makes his own design unusable. Neighbors within virtual cars must continuously communicate to realize the full potential of cooperative design.

4.1.2 Required IT System Environment

Collaborative creation and analysis of virtual cars requires a compatible and at least partly standardized IT system environment (see Fig. 4.1):

- *3D CAD modelers* such as CATIA, Pro/ENGINEER, NX or Solid Works are used to create the 3D models of vehicle parts. As almost all visible surfaces of a vehicle must meet very high aesthetic requirements and might initially have been modeled in clay (see Sect. 7.3.1.4), freeform surface modeling is among the most important capabilities of CAD systems in automotive design. Because of its strength in this domain, CATIA has become a quasi-standard in the automotive industry. Pro/ENGINEER is used especially in engine design, where the main surfaces are analytical rather than freeform, and the advantages of parametric design can be fully exploited (e.g. when designing a four-cylinder-inline and a six-cylinder inline engine as one family). Relevant capabilities of automotive 3D CAD systems include advanced surfacing, advanced solid modeling, the ability to handle large assemblies [1]. While with first generation 3D CAD systems designers had to select and adapt elementary geometries or surfaces to create part geometries, they today can use feature oriented modelers that apply their design intent (such as a bore hole pattern or a groove for a sealing ring) to the part. In the same way, current 3D CAD systems offer specialized tools to generate models of e.g. harnesses, tubes or pipes.
- *Product data management (PDM) systems* manage and track creation and change of all geometric and non-geometric product information such as part numbers, description, cost, material, technical drawings etc. and link it to the respective 3D models. To store and retrieve this data, PDM systems operate a product database. PDM systems manage different configurations, versions and variants of the product. Thus, PDM systems are the tools to manage building and maintenance of virtual cars, e.g. by creation of a virtual BOM, the hierarchical structure of the virtual vehicle. To visualize the product structure and facilitate selection of parts needed for specific investigations (e.g. visualization of all parts inside or partially inside a defined box), a *structure navigator* is used as a user interface for the PDM system (see Fig. 4.1).
- While CAD systems focus on the detailed creation and change of single parts, *3D visualizers* allow the user to envision bigger sub-assemblies or even a

whole vehicle. As typical operations are moving, rotating and zooming in the product in question, visualizers need simplified models of the distinct parts to reduce the required computational effort. For this purpose, an envelope model is tessellated for each part and linked to it by the PDM system. With a state-of-the-art visualizer, a complete virtual car can be visualized and moved in real time - depending of-course on the performance of the graphics computer used. Visualizers offer additional functionalities such as fly-through with or without section pane, kinematics simulation or distance checking (see Sect. 4.2.2).

For further functional evaluation (such as crash behavior or cabin temperature), specific simulation tools are used to analyze the geometric and non-geometric data contained in the virtual cars. Simulation of complete vehicle characteristics is discussed in detail in Chaps. 7 and 8.

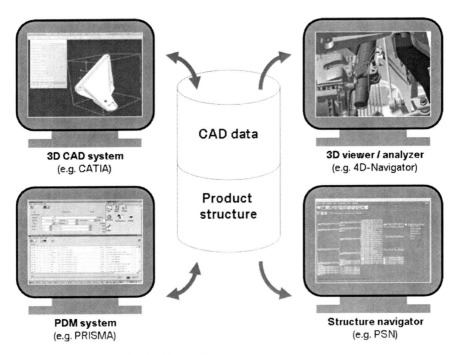

Fig. 4.1 IT environment for virtual car build

4.1.3 Specification

Starting with the concept phase, virtual vehicles are the shared workspace for co-operative complete vehicle design. Together with the integration processes, the *virtual car team* specifies the virtual vehicles regarding body type, country version, engine, gearbox and options. As provision and maintenance of virtual vehicles requires some effort, the key to virtual car specification is to find the least amount of vehicle configurations with which the required information can be obtained (compare Sect. 3.3.3). Figure 4.2 shows the structure tree for a set of specified virtual cars.

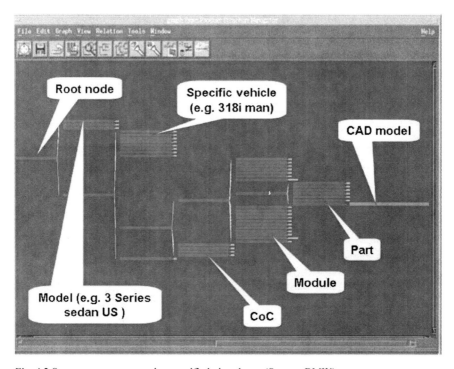

Fig. 4.2 Structure tree representing specified virtual cars (Source: BMW)

4.1.4 CA Data Provision

As a starting point for virtual cars, the virtual car team provides the structural environment and the shared workspace for each virtual car. Then, step by step the geometry data is added and positioned by the design CoCs. At the beginning of the concept phase, when the conceptual package of the new vehicle is created, components may be represented by simple geometric primitives (boxes, prisms etc.). The virtual car then becomes the basis for analytical design [2].

During the series development phase, every single part is separately modeled according to its geometric and functional needs. Over the PEP, the virtual cars grow and become more and more mature until their parts have a status that is ready for series production release [3]. Figure 4.3 depicts a detailed virtual car as it is used for series development.

Fig. 4.3 Detailed virtual car (Source: BMW)

4.2 Geometric Integration

The term *geometric integration* denotes all activities that ensure geometric coherence of the virtual vehicle throughout the PEP to prevent interferences and ensure provision of functional clearance.

4.2.1 Collision Detection

If two designers should both independently check their part changes in context before publishing it in a virtual car – and then the two changed parts interfere, or a designer does not check the complete environment of his part before publishing his new version; in concurrent engineering of virtual cars with its highly simultaneous data provision, clashes and interferences among virtual parts do occur and hence must be eliminated by geometric integration to ensure realizability of the vehicle.

Before geometric analysis can start, the quality and availability of CAD-models need to be verified: Are the models complete? Are the models positioned correctly? Are the models up-to-date? Figure 4.4 shows a virtual belt drive in two geometrically inconsistent scenes: One with the generator missing and one with a redundant obsolete belt.

Fig. 4.4 Redundant obsolete part (*left*) and missing part (*right*) (Source: BMW)

When the virtual car in question is complete and up-to-date, it is then automatically checked for collisions by tools that are part of the 3D visualizer. The identified collisions are communicated back to the designers responsible for the colliding parts so that they can evaluate the criticality of the collision [4]. Figure 4.5 shows such an automatically generated collision list with manually added criticality remarks. The colliding parts are specified in the two left columns, criticality is indicated by red or green color in the right column.

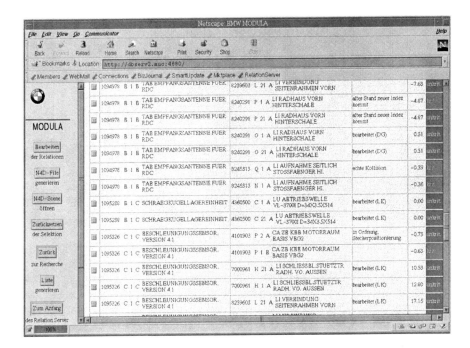

Fig. 4.5 Results from an automated collision check (Source: BMW)

As Fig. 4.6 illustrates, the criticality of the detected collisions among virtual parts can vary significantly:

- The collision in the left picture is uncritical. The detected interference between a rubber hose and a bracket is settled in reality by the elastic material properties of the hose. Another typical example for uncritical collisions is interferences between door seals and the body side frame. The seals are usually modeled in their un-deformed shape and positioned in relation to the door. When loaded into the complete vehicle representation, the seals naturally collide with the body frame model, because the virtual sealing does not deform as the real ones would do.

- The center picture of Fig. 4.6 shows a collision stemming from the misalignment between the bolts of a rear light assembly and the respective body holes. While the collision definitely has to be changed before the parts are released, change requires a rather minor effort: The holes can easily be shifted in the 3D CAD model. During the series production phase, this kind of collision happens frequently when one part's geometry is changed and the neighboring parts have not yet been updated accordingly.

- The collision shown in the right picture of Fig. 4.6 however is highly critical, an interference of rigid components. Concept-critical collisions like this require major design rework. During series development, this kind of full interference usually stems from wrong positioning of a part in the complete vehicle work space.

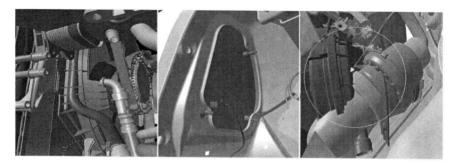

Fig. 4.6 Collisions: uncritical (*left*), minor (*center*) and critical (*right*) (Source: BMW)

Many designers feel that automated collision checks are a means to detect and publish design glitches and even feel provoked by the regular collision reports. They see collisions as normal, intermediate states during the product design process. "Wait for my next part version" is hence the typical measure that is proposed to solve their collisions. But as every designer can update his or her part at any time, it is very likely that the new version indeed removes the listed collision but at the same time creates another collision somewhere else in the virtual car. Automated collision checking is fundamental to design-in-context, which is the indispensable basis for simultaneous development of the complete vehicle.

4.2.2 Ensuring Functional Clearance

The term functional clearance denotes the minimum clearance between two parts that is required to ensure proper functionality. A minimum clearance can be required to avoid heat transfer (e.g. between an exhaust pipe and a thermal shield), to allow collision-free movement of moving parts in all degrees of freedom (e.g. movement of the front wheels in their wheel housings) or to give enough space for oscillatory movements (e.g. small rotations of the engine as it reacts to the driving torque). While interferences among virtual parts can easily be detected by means of collision checking tools, ensuring functional clearance is approached manually. 3D visualizers usually offer tools to measure the minimum distance between two parts. In the example in Fig. 4.7, the minimum distance between an engine bracket and a coolant hose routed below it is measured as 7.5 mm.

Fig. 4.7 Analysis of parts' clearance (Source: BMW)

The question of how much clearance is necessary can only be answered by the responding experts. *Clearance libraries* can provide initial recommended values. But as many parameters can influence the required distance, dependable values must be calculated or estimated by experts – e.g. the critical distance between a heater hose and the engine with regards to relative movement of the end fixed to the engine and the end fixed to the body, the elasticity of the hose which depends on the temperature and flow rate of the water, oscillations of the hose during operation of the vehicle etc.

4.3 Further Functional Geometry Evaluation

4.3.1 Storage of Personal Items

To ensure optimum practicality in terms of storing personal items in the cabin and the baggage compartment, these items are represented as 3D models and added to the virtual cars. Including this virtual load in the geometric integration processes, allows the size and accessibility of trays, pockets and the baggage compartment as a whole to be verified at an early stage of the PEP (compare Sect. 7.3.3.3).

4.3.2 Evaluation of Vehicle Kinematics

Another advanced application of geometric integration processes is the validation of vehicle kinematics. This includes:

- Operational movements such as shifting the gearshift lever in every position (with different kinds of bottles in the cup holder and the ashtray lid open or closed) or simply the movements of the doors, hood and trunk lid.
- Functional movements such as jouncing and rebounding of chassis and suspension parts and tires, turning-in of steering parts and front wheels or lowering and lifting the side windows.

An effective approach to check clearance for these movements is to create an envelope model for the moving part by sweeping the part's geometry along the trajectory of the movement and checking this envelope against the vehicle geometry (see [3]). An important application of this approach is the creation of the wheel envelope (see Fig. 4.8) which represents the maximum geometric expansion of all wheels and tires that should be used with the vehicle in question. Calculation of wheel envelopes must take into account the following parameters:

- Maximum jounce and rebound allowed by the suspension kinematics
- Maximum turn and tilt allowed by the steering kinematics
- Maximum elastic deformation of the tire according to longitudinal, vertical and lateral forces
- Maximum elevation of snow-chains (if applicable)
- Maximum and minimum tire pressure
- Wear conditions of the tire

This envelope is then checked for clearance against the wheel housing and components such as brake hoses, ABS sensor cables, or handbrake bowden cables. But the particular relevance of the front wheel envelope stems from its role in determining the geometry of the concept vehicle: During the concept phase, the front wheel envelope determines the maximum outer limits of the engine cradle and thus – with the width of the engine and the engine cradle beams given – the total width of the vehicle (compare Sect. 3.1.1).

Fig. 4.8 Wheel envelope as used for concept evaluation (Source: BMW)

Geometric evaluation of complex kinematics such as that of a retractable hard top is not possible with this quasi-dynamic envelope approach. It requires advanced multi-body simulation systems, that analyze the kinematics of parts linked together by joint-functions that constrain their relative movements [5].

4.4 Virtual Build Groups

Parallel to the "regular" continuous vehicle development, certain project milestones require the validation of the status of the complete vehicle at a specific point in time. While this was formerly done by means of hardware prototype build groups, today *virtual build groups* are carried through to create a comprehensive picture of a vehicle's maturity. Virtual build groups can be structured in three main phases:

- During the *planning phase*, the vehicle properties which must be validated are agreed. According to these expected results, the required virtual cars are then specified by a *build group specification*. Usually, the virtual cars of a build group are copies of the existing virtual cars.
- In the *design phase*, supporting this design-in-context, the virtual car team iteratively checks geometric compliance of the virtual cars (geometric integration). Geometric problems are evaluated and fed into a problem management

process (see Sect. 6.2) to monitor their solution. At the end of design phase, CAD and CAE data are 100% consistent: Complete, correct, clash-free. The virtual car is geometrically released and ready to be tested.

- During the *evaluation phase*, functional integration and production integration test functions and buildability of the virtual vehicles using the appropriate simulation tools and methods. Eventually, the findings of the virtual build group are reported to management to formally verify the project status.

The first two phases of a virtual build group are also carried through to ensure CAD data maturity for hardware prototypes before ordering prototype tooling and parts.

References

1.   Gould LS (2008) What makes automotive CAD/CAM systems so special? Automotive Design and Production. http://www.autofieldguide.com/articles/109802.html. Accessed 20 November 2008
2.   Braess HH, Seiffert U (2005) Handbook of automotive engineering. SAE, Warrendale (PA)
3.   Kelley DS (2007) CATIA for design and engineering. Version 5. Releases 14 & 15. SDC Publications, Mission (KS)
4.   Weber J (1998) Ein Ansatz zur Bewertung von Entwicklungsergebnissen in virtuellen Szenarien. Institut für Werkzeugmaschinen und Betriebstechnik, Universität Karlsruhe
5.   Zamani NG, Weaver JM (2007) CATIA V5 Tutorials. Mechanism design and animation. Release 16. SDC Publications, Mission (KS)

Chapter 5
E/E System Development

Abstract Conventional vehicle development was mainly about designing and testing mechanical and electro-mechanical components. But over the last decades, cars have become complex systems of E/E systems (electrical and electronic systems), and electronics and software development have been integrated into automotive development processes. This chapter describes both the technical structure of automotive E/E systems and their development processes.

5.1 From Machinery to E/E Systems

5.1.1 A New and Different World

About 7 years ago, leading OEMs launched their luxury class vehicles loaded with the latest electronic features. Some of these vehicles – much to the surprise and annoyance of both their customers and their manufacturers – experienced random and unpredictable electronic malfunctions. Manufacturers with an international reputation for quality and reliability suddenly struggled to deliver cars that worked the way they were supposed to work [1].

What happened? In this generation of cars, almost all functions were electronically controlled – and also interlinked. Together with the increasing number of variants, this led to a quantum leap in the complexity of the resulting overall E/E system – to a degree that the established vehicle development processes that had been honed over decades to efficiently create high quality mechanical systems were not able to cope with. Automobiles had changed from machinery to systems of E/E systems. The following facts make this paradigm shift evident:

- Cars have up to 2500 software controlled functions, representing 10 million lines of software code
- These functions are realized by up to 80 ECUs (electronic control units) that communicate via up to 5 different types of system busses
- Up to 40% of a vehicles' costs are determined by electronics and software
- 90% of all innovations are enabled by electronics and software
- 50–70% of the development costs for an ECU are related to software

J. Weber, *Automotive Development Processes*, DOI 10.1007/978-3-642-01253-2_5,
© Springer-Verlag Berlin Heidelberg 2009

For the future, it can be expected that E/E hardware (such as ECUs, sensors etc.), operating system software, and interfaces (see Sect. 5.2.5.1) will be more and more standardized, even over multiple OEMs. The differentiation between different brands and manufacturers will be mainly based on the proprietary applications software.

5.1.2 Automotive E/E Systems

In very simple terms, an electronic vehicle function is always realized by an actuator that e.g. moves, heats up, illuminates, or displays something, according to input generated by a sensor or given by a human user, and controlled by a software program that is embedded in an ECU. A complete vehicle's E/E system hence consists of all the vehicle's components that carry electric current, which are:

- Sensors (e.g. wheel speed sensor)
- Input devices (e.g. push buttons, rotary selector)
- ECUs with embedded control software
- Actuators (e.g. motors, lights, heating elements)
- Displays and speakers
- Harnesses for data and power (cables and connectors)
- Battery
- Generator/alternator

Among these elements, embedded software plays an increasingly dominant role. Today, up to 70% of the costs for an ECU are related to software. The head unit of the new 2009 BMW 7 series e.g. includes 4 million lines of code.

Figure 5.1 illustrates the complexity of modern vehicle E/E systems, taking the current BMW 7 series as an example.

Automotive E/E systems are divided into clusters of related functions[9] – so-called vehicle *domains*. The following list represents a domain structure that is widely used in the automotive industry:

- Infotainment: Functions for information, entertainment and communication (e.g. instrument panel, audio system, antenna tuner, video module, navigation system, telephone etc.), including the central HMI (Human Machne Interface) (compare Sect. 7.5)
- Body electronics: Central vehicle functions such as access management systems and anti-theft devices (compare Sect. 7.8), actuation of windows, tailgates, seats, and wipers, or cabin air management (compare Sect. 7.4.3)

---

[9] Often, the technical system architecture reflects the structure of the development organization, and each domain includes exactly those functions which are handled by one specific department.

- Chassis and driver assistance: Functions enabling safe vehicle dynamics, such as ABS, ESP, ASC, DSC (compare Sect. 7.6)
- Powertrain: All functions controlling the generation of driving power and its conversion into propulsion: Digital engine control, gear control, fuel pump, on-board diagnostics etc.
- Passive safety: Functions for collision detection and injury mitigation such as seat belt pre-tensioners, airbags, pyrotechnic roll-over bars etc. (compare Sect. 7.7)

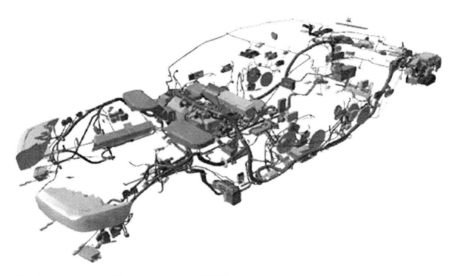

Fig. 5.1 Automotive E/E system (Source: BMW)

Components that belong to one domain exchange data over a common bus system. Ideally, the organizational structure of E/E the development follows the technical domain structure, even if in some cases the domain structure represents the organizational structure.

A *sub-system* comprises all the components that are required to fulfill one or more specific functions. This can range from only a few components to almost the whole system. As an example, Fig. 5.2 shows the components of the car access control sub-system.

In parallel to the data network, the E/E system also requires a power network that supplies all electric loads with the required standard power of 12 V DC (with the exception of ignition and electric drive motors).

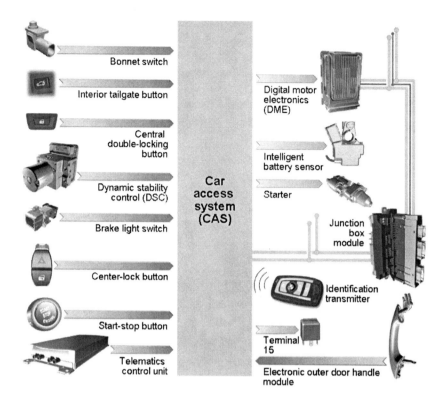

Fig. 5.2 Components of the car access control sub-system (Source: BMW)

5.2 Systems Engineering Processes

5.2.1 A Clash of Cultures

When considering the reasons why the previous vehicle generation experienced quality problems with the automotive E/E systems discussed above, one sees that one major reason for OEMs' and suppliers' development processes not keeping pace with the increasing product complexity actually was a clash of cultures between the "old economy" and the "new economy".

Automotive development unquestionably has its roots in traditional engineering, and the prevailing culture in development centers is dominated by "car guys" – mostly with a background in mechanical, electrical or control engineering – and their idea of how product development should work. But with the rapid increase of electronically controlled functions in cars over the last decade, a different breed of experts have almost unnoticedly populated development centers: Computer scientists and software engineers, who mostly came from non-automotive and even non-industry companies, and apart from wearing different clothes and living different lifestyles – which actually made integration on a personal level difficult – had different ideas of (and needs for) product development, complete vehicle integration or quality and reliability (see Sect. 7.9.1 on the different definitions of reliability). This adverse understanding is very nicely illustrated by the famous – but fictitious – controversy between Bill Gates and the chairman of General Motors, in which Gates complains about the low performance of vehicle development compared to software development, and in return the chairman of GM pictures what the owner of a car would experience, if the car satisfied the same quality and reliability standards as Microsoft's software does. While engineers developing chassis or body parts consistently checked with geometric integration to make sure that the complete vehicle remained geometrically coherent, software engineers worked more or less independently from each other and without a stringent process for the functional integration of the complete system. Spontaneous changes to a program that were meant to fix or improve certain functions could untraceably lead to malfunctions somewhere else and made problem analysis on a complete vehicle level an almost Sisyphean challenge.

5.2.2 Systems Engineering

Out of this situation, OEMs introduced systems engineering as a sustainable approach to developing reliable E/E systems as part of the PEP. The International Council on Systems Engineering (INCOSE) defines System Engineering as an interdisciplinary means to enable the realization of successful systems by (in this sequence) early focus on customer needs, documenting requirements, design synthesis, and system validation while considering all aspects of the posed problem [2]. The V-model in Fig. 5.3 lists the main elements of the systems engineering process, which will be discussed in detail in the subsequent subsections.

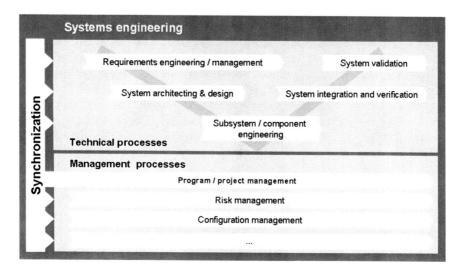

Fig. 5.3 System management V-Model (Source: BMW [3])

5.2.3 Requirements Engineering

All studies on why system development projects fail name incomplete clarification of customer requirements as one of the top reasons. A structured collection and analysis of requirements and a management process that keeps requirements up-to-date over the course of the project are an indispensable basic prerequisite for successful system realization. Requirements engineering follows four steps: Elicitation, analysis, specification, and validation.

5.2.3.1 Requirements Elicitation

As a basic principle of the elicitation process it must be ensured that all relevant stakeholders' requirements are included from the very beginning. Thus, the first step in requirements elicitation is identification of the relevant stakeholders of the system. Stakeholders do not only include all different types of direct customers who operate the system (primary customers), but also all indirect or secondary customers along the life cycle (development, manufacturing, deployment, support, disposal, training, and verification) who have their individual and often conflicting requirements regarding the system [4].

The second step is to actually gather the concrete requirements from the identified – human – stakeholders. Here, the primary challenge is to translate the requirements from the stakeholder's different languages and conceptual models into a unified description, which requires sensitivity and good understanding of the stake-holders' worldview by the system engineer. Typical elicitation methods include:

- Stakeholder interviews
- Observation of stakeholders
- Testing and commenting on the usability of a system prototype
- Testing of existing systems

5.2.3.2 Requirements Analysis

While the result of the requirements' elicitation is more or less a raw collection of untuned ideas, wishes and desires, the result of the requirements analysis step is a list of the functions the system should be able to perform, including how well the system should be able to perform theses functions, and with all issues resolved concerning [5]:

- Possible conflicts
- Interfaces to the environment
- Completeness
- Ambiguity

The first classification of requirements is typically made between functional and non-functional requirements, according to the type of the respective stakeholder. Functional requirements describe the future system's functionality from the user's (the primary customer's) point of view. Non-functional requirements do not directly support customer-relevant functionality but still must be met by the system. Table 5.1 lists up sample requirements for the ECU of a convertible roof system:

Table 5.1 Sample requirements for a convertible roof system (Source: BMW)

Requirement	Type
"Opening and closing of the roof must be possible by means of a switch in the vehicle's interior"	Functional
"If parked in the rain, the roof should automatically close"	Functional
"Opening or closing of the roof must not be possible at vehicle speeds over 30 km/h."	Non-functional
"Flashing the roof system controller may require 5 minutes max."	Non-functional

In order to serve as a robust basis for the future vehicle, system requirements must meet distinct quality criteria. According to [4], they must be:

- Achievable
- Verifiable
- Unambiguous
- Complete
- Expressed in terms of need, not solution
- Consistent with other requirements
- Appropriate for the level of system hierarchy

5.2.3.3 Requirements Documentation and Validation

Eventually, system requirements must be specified in a formal requirements specification document that can be systematically reviewed, evaluated, and approved, and which serves as the binding basis for the following system design. Typical forms of documentation are use cases (specifications of how the system should respond to a certain request from outside) or process specifications. IT tools that support requirements management (e.g. DOORS from Telelogic/IBM) usually use database systems that not only allow structured documentation but also attribution and linkage of requirements, and offer an interface to general office applications.

5.2.4 System Architecture and Design

System design starts with the creation of an architecture derived top-down from the specified requirements. While the requirements specification describes WHAT the vehicle's E/E system should do, the system architecture describes HOW the system should do it – both on a logical/functional and a technical level. As it defines the vehicle's interface to components that are usually developed and manufactured by suppliers, creation of the system architecture is a core competency of an automotive OEM – or a highly specialized development partner.

5.2.4.1 Logical System Architecture

The system level requirements are decomposed into sub-functions which together with their internal and external interfaces represent the *logical system architecture*. While the internal interfaces describe the required communication lines between the distinct sub-systems, the external interfaces describe inputs from and outputs to the system's environment, especially the driver and passengers. The logical system architecture is represented by a block diagram in which each sub-function is represented by one block. As an example, Fig. 5.4 shows such a block diagram for the BMW Night Vision system.

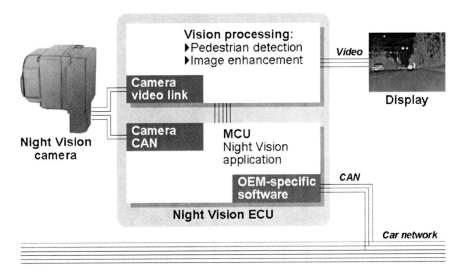

Fig. 5.4 Block diagram for a sub-system (Source: Autoliv)

5.2.4.2 Technical System Architecture

Partitioning

In the next step, the so-called partitioning process, the functions of the logical system architecture are mapped to the realizable elements of the technical system architecture (see Fig. 5.5): Distinct ECUs, sensors, actuators, bus systems (hardware architecture) and application programs (software architecture). The mapping of software elements (application programs) to distinct hardware elements (ECUs) is referred to as deployment.

The increasing performance of mobile internet connections has enabled specific vehicle functions to increasingly be partitioned to computers outside of the vehicle, so-called back-ends. In this case, the vehicle actually runs web applications (e.g. an internet based route planning algorithm for the navigation system) instead of on-board software.

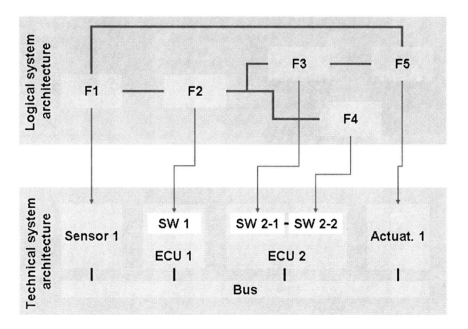

Fig. 5.5 Illustration of the partitioning process

In order to optimize the system at an early stage, alternative partitions are created and rated regarding the following criteria:

- Degree of requirements' fulfillment
- Technology risk
- Development risk
- Process aspects
- System performance
- Organization
- Compatibility
- Longevity

Eventually, the optimum partition with the best overall rating is selected for development.

Bus Systems

In parallel to the distribution of functions to specific hardware and software components, creation of the technical system structure also includes the specification of the external bus systems used for inter-ECU-communication. A bus system is specified by the protocol, the technical properties of the carrier (e.g. cable), and the topology according to which the distinct ECUs are interlinked. Which bus system is the most suitable depends on the domain's prevailing requirements regarding speed, volume and reliability of data transfer:

- Infotainment: While multimedia functions use extremely high transmission rates (e.g. video streams), requirements regarding reliability are comparatively low. Typical bus protocols are MOST or Ethernet.
- Body electronics: With requirements typically being high regarding reliability (e.g. car access, wiper etc.) and low regarding data rates, the typical bus protocol is low speed CAN.
- Chassis and driver assistance: As these functions require reliable real-time response, the typical bus protocol is high speed CAN.
- Powertrain: Data transfer requirements are similar to those of chassis and driver environment. Hence, the typical bus protocol is also high speed CAN.
- Passive safety: Requires the highest possible degree of reliability and real-time response. Typical bus protocols are TT CAN, byteflight or FlexRay.

Taking the current BMW 7 series as an example, Table 5.2 compares the specifications of the bus systems used in the different vehicle domains.

Table 5.2 Technical specifications of the E/E Domains in a luxury class vehicle (Source: BMW)

	Infotainment	Body electronics	Chassis and Driver assistance	Powertrain	Passive safety
Program size	100 MB	2,5 MB	4,5 MB	2 MB	1,5 MB
ECUs (standard)	4	14	6	3	11
ECUs (options)	12	30	10	6	12
Bus type	MOST	K-CAN	F-CAN / PT-CAN	LoCAN / PT-CAN	SI-BUS
Bandwidth	22 MBit/s	100 KBit/s	500 KBit/s	500 KBit/s	10 MBit/s
Messages	660	300	180	36	20
Cycle time	20 ms–5 s	50 ms–2 s	10 ms–1 s	10 ms–10 s	50 ms
Reliability requirements	Low	High / low	High	High	Very high
Transmission medium	Optic fiber	Copper cable	Copper cable	Copper cable	Optic fiber
Bus topology	Ring	Tree	Tree	Tree	Star

Figure 5.6 shows the complete technical system architecture of the same vehicle. The buses connecting the components of the separate vehicle domains are linked to each other by means of a central gateway module which also steers the connection to off-board computers and data storage via wireless connections.

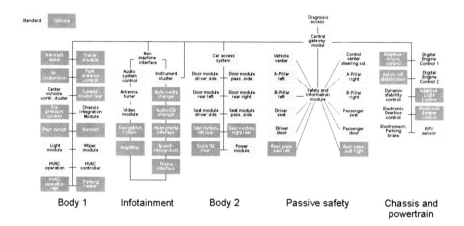

Fig. 5.6 Complete vehicle technical system architecture of luxury class vehicle (Source: BMW)

Not only must the technical systems architecture represent a working E/E system; it also must be integrated into the complete vehicle development – e.g. packaging (compare Sect. 3.2) and geometric integration (compare Sect. 4.2), production integration (compare Sect. 8.1), or service integration (compare Sect. 8.2).

The different steps of system architecture design are supported by a variety of methods and IT tools. *Function nets* as the standard representation of logic system architectures are modeled in UML-RT which has become a quasi-standard modeling technique used by various IT tools such as e.g. Rational Rose Technical Developer from IBM.

5.2.5 Component Development

5.2.5.1 ECU Hardware and Embedded Software

In the context of E/E systems development, component development denotes the design of the distinct ECUs and other E/E elements (sensors, actuators etc.) and validation of the resulting sub-system. The main elements of an ECU are:

- Micro controller. A functional computer built in one chip that includes CPU, RAM, EEPROM, I/O etc.

- ECU basic software, including operating system, internal and external data communication, network management etc., e.g. according to OSEK / ISO 17356
- Application software: The program that controls and executes the required functions.
- Middleware that allows application programs to use a common API (application programming interface) which is independent from both the data bus system and the operating system.
- Mechanical elements: printed circuit board (PCB), housing, shielding, terminal sockets etc.

Middleware is the critical prerequisite for exchangeability of software and hardware components among model lines, OEMs and/or suppliers. Here, the AUTOSAR run time environment has become an automotive industry standard.

By specifying interfaces and their communication mechanisms, the application programs are decoupled from the underlying hardware and basic software, enabling the realization of standard library functions [6]. Figure 5.7 shows the AUTOSAR ECU software architecture.

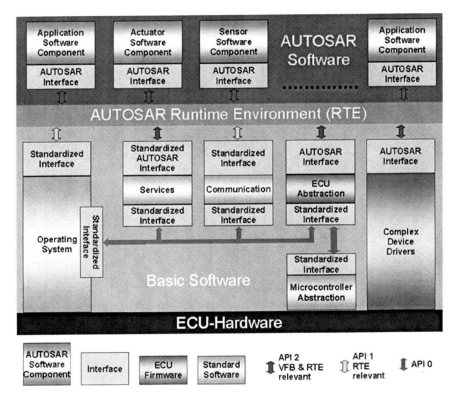

Fig. 5.7 AUTOSAR ECU software architecture (Source: AUTOSAR)

Design of the ECU hardware includes dimensioning and selection of the required electronic component parts, circuit design, PCB layout, specification and placement of the required interfaces and design of appropriate component housing (see e.g. the engine controller (front) and Valvetronic controller (rear) in Fig. 5.8). The respective mechanical tests are discussed in Sect. 7.9.3.3.

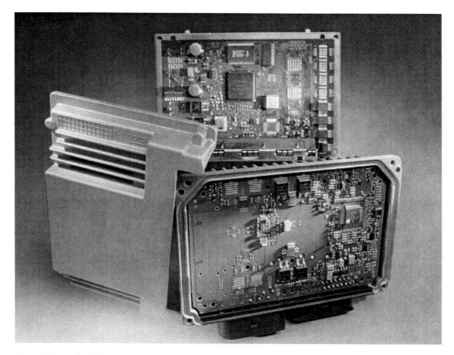

Fig. 5.8 Sample ECU hardware (Source: BMW)

5.2.5.2 Application Software Design

The core process in component design is the generation of the application software which defines the components' behaviour. As a first step, a function model is created that specifies the required mathematical functions, algorithms and the component interfaces (inputs and outputs). IT tool kits (e.g. Matlab from The Math Works or ASCET from ETAS) offer comprehensive support for functional modelling, MIL (model-in-the-loop) and SIL (software-in-the-loop) simulation (in which the complete hardware environment is emulated) and problem analysis down to the generation of the application software's C code – which is compiled to assembler code.

Then, in the application process, parameters of the software (e.g. the engine and gearbox type as parameters of the engine controller software) are set. This application data, the assembler program and additional data required e.g. for diagnosis is subsequently linked to a binary program file which then is flashed ("embedded") in the controllers EEPROM.

5.2.5.3 Component Testing

The next step in E/E component design is testing of the ECU with the embedded application software. To allow parallel development of the ECU and its peripheral components, these tests are carried out in HIL (hardware-in-the-loop) test rigs, which electronically emulate the ECU's interface (sensors and actuators). A HIL simulation platform for an instrument panel e.g. (see Fig. 5.9) emulates all of the sensors that deliver the data required for the displayed information (e.g. engine speed, vehicle speed, fuel level etc.).

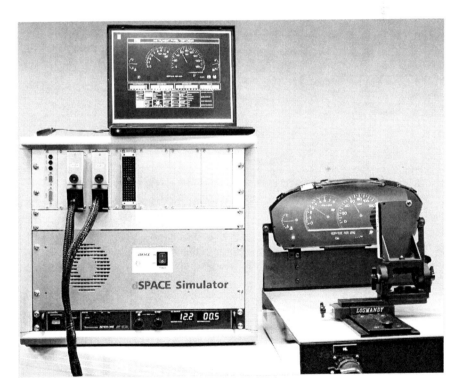

Fig. 5.9 HIL test bench for an instrument panel (Source: dSPACE)

5.2.6 Systems Integration and Validation

System integration means to physically and functionally bring together the components that have been developed by the respective (internal or external) design CoCs into one complete vehicle system and to ensure that this system fulfills the desired functions. Over the PEP, a formal system integration process (see Fig. 5.10) is carried through before each build phase (prototype build group 1, prototype build group 2, pre-series production etc. – compare Sect. 3.3) in order to ensure the respective vehicle's functionality.

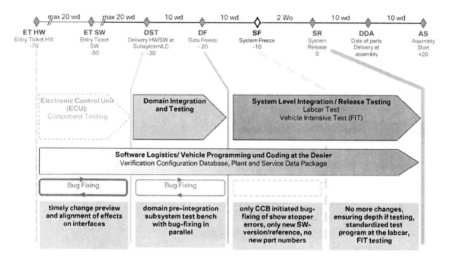

Fig. 5.10 E/E systems integration process (Source: BMW)

At the beginning of this process, each component a design CoC wishes to be a part of the vehicle's E/E system must be registered for system integration. After being pre-qualified by the CoC and handed over to the systems integration department, the respective component is under the strict change control of the system integration process where it is evaluated as part of a sub-system or complete vehicle system.

5.2.6.1 Sub-system Integration

On a sub-system test bench, all components which are required to fulfill a certain function are connected by the appropriate bus systems. In order to create realistic behavior of the sub-system under test, components that are not part of the sub-system are simulated. Sub-system level tests include customer and system functions, e.g. mechanical, chemical or atmospheric strain, quiescent current, electro-magnetic compatibility etc. As an example, Fig. 5.11 shows a sub-system test bench for the vehicle light system.

Fig. 5.11 Vehicle lights sub-system test bench (Source: BMW)

Test runs on subsystems include:

- Coding
- Flashing
- Diagnosis jobs
- Bus diagnosis (CAN DTCs)
- Sleep and wakeup tests
- Selected user/customer functions

Problems found during the sub-system tests are fed back to and subsequently addressed by the design CoCs until the sub-system under test is validated.

5.2.6.2 Complete Vehicle Integration

After the sub-systems have been separately built, tested and validated, they are consecutively integrated to create the complete vehicle E/E system. The distinct sub-systems are physically and functionally linked together to test their correct interaction and fulfillment of the system level functions. As a first step, all of a vehicle's E/E components are connected and laid out in a stationary lab-car (see Fig. 5.12), where system functions and non-driving customer functions are evaluated.

Fig. 5.12 Stationary lab-car (Source: BMW)

Tests on stationary lab cars include:

- Pinning / wire harness
- Bus physics
- Power up / power down
- Voltage tests
- Selected user/customer functions.
- Sleep current
- 24 h test
- Bus load
- Misuse test
- Gateway test

- Coding
- Flashing
- Diagnosis jobs

A stationary lab-car includes about 80% of a complete vehicle's E/E system. To evaluate the whole system from its internal and external customers' view, complete vehicles from the respective build group must be investigated. These so-called *dynamic lab-car*s are furnished with comprehensive measuring equipment (see Fig. 5.13) that documents all data bus traffic and thus allows tracking electronic malfunctions and validating of dynamic E/E functions such as engine control, gearbox control or chassis control systems. Again, the detected problems are fixed by the design CoCs.

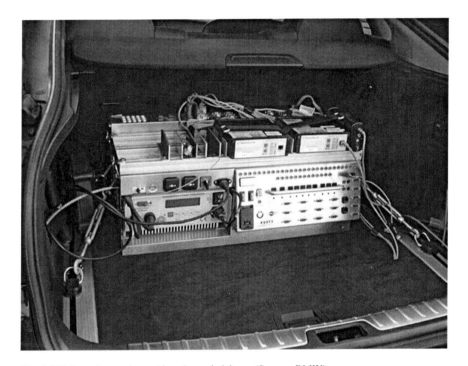

Fig. 5.13 Measuring equipment in a dynamic lab-car (Source: BMW)

At the end of the systems engineering process, systems integration releases a defined and verified configuration of the complete vehicle E/E system, a so-called *integration level* (see below).

5.2.7 Supporting Management Processes

Alongside the technical development processes already discussed, management processes support successful realization of the desired E/E system in the complete vehicle. In addition to general project management, the two major supporting management processes are *configuration management* and *risk management*.

5.2.7.1 Configuration Management

Just as geometric integration (see Sect. 4.4) ensures geometric consistency of the complete vehicle's design prior to a hardware build group (see Sect. 3.3.3), system integration leads to valid E/E system configurations – functionally consistent sets of all of a vehicle's hard- and software. Therefore, the systems engineering process according to the V-model shown in Fig. 5.3 is reiterated before each of a vehicle project's hardware build phases (see Fig. 5.14).

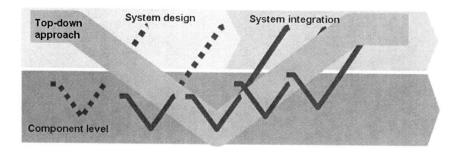

Fig. 5.14 Integration and verification loops over the PEP (Source: BMW [7])

Here, the task of configuration management is to structure, label, document, and control all versions and variants of soft- and hardware components as well as all valid system configurations made up by them.

Changes to software or hardware artifacts must not be done ad hoc, but controlled by a change control board (CCB) that assesses change requests and analyzes their potential benefits and risks for the complete system.

In order to realize the required E/E functionality in the hardware vehicles, configuration management also maps the valid system configurations to the respective vehicle variants (e.g. engine variants, country specific variants, options etc.), and ensures that only the latest validated configurations are built into the vehicles.

Figure 5.15 shows the configuration planning for one vehicle project. The system configuration of the first vehicles of the first prototype build group (grey) is denoted as integration level I200, subsequent updates as I210, I220 etc. Analogously, integration levels for the second prototype build group (blue) are called

I3xx, for pre-series vehicles (yellow) I4xx and for the series vehicles that are sold to customers (green) I5xx.

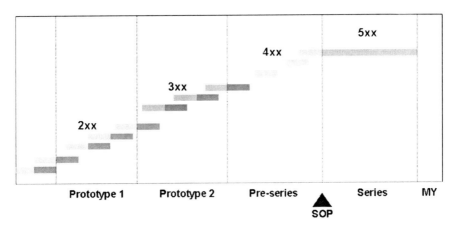

Fig. 5.15 System configuration planning (Source: BMW)

5.2.7.2 Risk Management

In each vehicle project, there is always a multitude of things that could happen during the development of the E/E system that would have a negative effect on the planned achievement of system objectives and performance in terms of quality, schedule or cost. Independent from whether engineers and managers are aware of them, acknowledge them or plan for them, these risks exist and knowing them is an opportunity to avoid possible resulting problems [4].

The prerequisite for a risk is a future root-cause, which, if eliminated or corrected, would prevent a potential negative consequence from occurring. Hence, the two parameters that specify a risk are:

- The probability of occurrence of that future root-cause
- The consequences of the occurrence of that root-cause

Risk management as an essential support process of systems engineering includes four main elements [4]:

- Planning of objectives, activities, resources qualification, communication etc.
- Risk identification and quantitative assessment. Risks can be related to the E/E system itself including upstream external systems, or to the system development processes and their external influences.
- Implementation of risk mitigating or eliminating measures. This includes risk avoidance (e.g. through trading risks for system performance), lowering the risk to acceptable levels through system design measures, acceptance of the

risk, or transferring the risk from one system area to another (e.g. from software to hardware).

- Monitoring, documenting and reporting the measures taken and thus ensuring successful risk mitigation, re-assessment of the risk situation and communication to management.

Figure 5.16 illustrates how the four elements of risk management are in mesh:

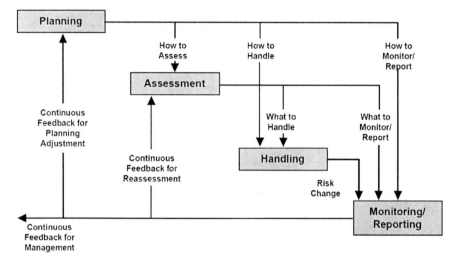

Fig. 5.16 Risk management control and feedback
(Source: U.S. Department of Defense, Systems Management College [4])

5.2.8 CMMI

Successful realization of a vehicle's E/E system – on schedule and within budget – as part of the PEP largely depends on the quality of the established system development processes. The *Capability Maturity Model Integration* (CMMI) is a set of process models developed by the *Carnegie Mellon Software Engineering Institute* (SEI) that provides goals (proven best practices) for the effective organization of software or systems engineering processes. These goals are specified by 22 process areas (see Table 5.3), clusters of related practices that, when implemented collectively, satisfy a set of goals considered important to improve in that area. The process areas are clustered into four categories [8].

In terms of process assessment and improvement, CMMI methodology offers two different approaches: A staged representation that rates the organizational maturity, and a continuous representation of the capability an organization has in the distinct process areas.

Table 5.3 CMMI process areas
(Source: Carnegie Mellon University, Software Engineering Institute [8])

Category	Process area
Project management	- Project planning
	- Project monitoring and control
	- Supplier agreement management
	- Integrated project management
	- Integrated supplier Management
	- Integrated teaming
	- Risk management quantitative project management
Support	- Configuration management
	- Process and product quality assurance
	- Measurement and analysis
	- Causal analysis and resolution
	- Decision analysis and resolution
Engineering	- Requirements management
	- Requirements development
	- Technical solution
	- Product integration
	- Verification
	- Validation
Process management	- Organizational process focus
	- Organizational process definition
	- Organizational training
	- Organizational process performance
	- Organizational innovation and deployment

5.2.8.1 Organization Maturity Levels

Focusing on software development rather than on systems development, the staged representation splits the process of establishing the best practices in the 22 process areas into five steps that represent increasing organizational maturity levels (from 1 to 5). Reaching one maturity level denotes the successful implementation of best practices for a specific subset of the 22 process areas (see Table 5.4) and is a prerequisite for implementing the processes required for the next higher maturity level. An organization at maturity level 5 has achieved all the goals in all 22 process areas.

Table 5.4 CMMI maturity levels
(Source: Carnegie Mellon University, Software Engineering Institute [8])

Level	Focus	Key process area
Level 1 Initial	Process is informal and ad-hoc	
Level 2 Managed	Basic project management	- Requirements management - Project planning - Project monitoring and control - Supplier agreement management - Measurement and analysis - Process and product quality assurance - Configuration management
Level 3 Defined	Process standardization	- Requirements development - Technical solution - Product integration - Verification - Validation - Organizational process focus - Organizational process definition - Organizational training - Integrated project management - Risk management - Decision analysis and resolution - Integrated teaming - Org. environment for Integration - Integrated supplier management
Level 4 Quantitatively Managed	Quantitatively managed	- Organizational process performance - Quantitative project management
Level 5 Optimizing	Continuous process improvement	- Organizational innovation and deployment - Causal analysis and resolution

5.2.8.2 Process Capability Levels

While the maturity levels' focus on the stepwise improvement of a development organization as a whole, CMMI capability levels apply to an organization's incremental process improvement in individual process areas. Table 5.5 lists the six increasing levels of process capability, numbered 0 through 5:

Table 5.5 CMMI capability levels
(Source: Carnegie Mellon University, Software Engineering Institute [8])

Capability level	Process characteristics
Level 0: Incomplete	Process is not or only partially performed. At least one goal specified for the process area is not satisfied.
Level 1: Performed	Process satisfies the goals of the process area. Process supports and enables the work needed to produce the required results. Capabilities can be lost over time, if not institutionalized through generic practices (capability levels 2 through 5).
Level 2: Managed	In addition to meeting the requirements for level 1, the process is supported by a basic infrastructure, is planned and executed in accordance with policy, employs skilled people who have adequate resources to produce controlled outputs, involves relevant stakeholders, is monitored, controlled, and reviewed, and is evaluated for adherence to its process description. Level 2 capabilities help retaining existing practices even under stress.
Level 3: Defined	In addition to meeting the requirements for level 2, the process is tailored according to the organization's guidelines, and contributes work products, measures, and other process improvement information to the organizational process assets. More rigorously specified than at level 2, level 3 states purpose, inputs, entry criteria, activities, roles, measures, verification steps, outputs, and exit criteria.
Level 4: Quantitatively managed	In addition to meeting the requirements for level 3, the process is statistically controlled, and managed using quantitative quality and performance objectives.
Level 5: Optimizing	In addition to meeting the requirements for level 4, the causes of process variation are fully understood and used as the basis of continuous process performance improvement.

References

1. Reichle J (2003) Mehr Technik, mehr Pannen. Süddeutsche Zeitung vom 10.09.2003, Wirtschaftsteil
2. INCOSE (2007) Systems engineering handbook. INCOSE, Seattle (WA)
3. Negele H, Schmidt R, Finkel S, Wenzel S (2006) Lessons learned from synchronizing complex systems development within automotive industry. Proceedings of INCOSE symposium. 8–14 July 2006, Orlando (FL)
4. USDOD (2001) System engineering fundamentals. Defense Acquisition University Press, Fort Belvoir (VI)
5. IEEE (2004) Guide to the software engineering body of knowledge. IEEE, Washington (DC)
6. AUTOSAR Partnership (2006) Achievements and exploitation of the AUTOSAR development partnership. Convergence 2006. Detroit (MI), October 16 2006

7. Negele H (2006) Systems engineering challenges and solutions from an automotive perspective. INCOSE International Symposium 2007. 27 June 2006, Orlando (FL)
8. Software Engineering Institute (2006) CMMI for development, Version 1.2. Improving processes for better products. Carnegie Mellon University, Pittsburg (PA)

Chapter 6
Management Processes for Complete Vehicle Development

Abstract   To fully control the progress of a vehicle development project and optimally steer the resources involved over the course of the PEP, project management employs a variety of methods and processes. *Target management* e.g. includes the derivation, co-ordination and update of the technical vehicle targets; *problem management* deals with registration and control of vehicle design issues detected by complete vehicle integration; *release and change management* comprises unambiguous notation of component releases and evaluation of design changes; *quality management* methods ensure over all vehicle development and production activities such that project and product reach the intended performance goals.

6.1 Target Management

6.1.1 Complete Vehicle Requirements

One of the important trivia of product development is that the development result should meet the requirements. Complete vehicle requirements (compare Sect. 5.2.3) are the total of *legal requirements* that must be met to allow registration of the vehicle in the intended markets and *customer requirements* that denote direct perceptions of the vehicle's functions and characteristics by the individual users. While legal requirements are usually well-defined by governmental directives and regulations (such as FMVSS for safety standards in the U.S.), customer requirements are rather subjective. Customers seek vehicle functions and characteristics that meet their personal requirements – of course in addition to all legal requirements.

The different areas of customer relevant functions and characteristics – such as design appearance, dynamics or cabin comfort – and the respective legal or individual requirements are discussed in detail in Chap. 7, together with the processes and methods necessary to eventually realize these functions and characteristics in the complete vehicle.

J. Weber, *Automotive Development Processes*, DOI 10.1007/978-3-642-01253-2_6,
© Springer-Verlag Berlin Heidelberg 2009

One prerequisite for successful vehicle design is the necessary differentiation between requirements (what the customer wants or needs) and technical functions (means by which the engineer seeks to solve the problem posed by the customers' requirements) [1]. The relevant issues for the customer define the *customer domain*. Using thermal comfort as an example, driver and passengers simply want to have an agreeable climate in the vehicle cabin as indicated by e.g.:

- Perceived temperature at head / body / feet
- Perceived temperature of interior surfaces, especially seat covers
- Perceived temperature stability
- Perceived air flow
- Time needed for window defrosting
- Ease of temperature and air flow adjustment
- Additional fuel consumption caused by the heating, ventilation and air condition (HVAC) unit

Designers however formalize these needs in the *functional domain* by functional requirements and further more in the *physical domain* by design parameters of released parts and components – which are both of only secondary interest to the customer. In the thermal comfort example above, design parameters would be e.g.:

- Measured air temperature at left/right outlet
- Measured flow rate through left/right air ducts
- HVAC compressor power
- Seat heater power

When the different domains are confused, poor solutions may result. A designer who provides a functionality to adjust the vehicle clock should not forget that the actual requirement of the customer is not to adjust the clock but to have the correct time on display. Hence, the right solution might rather be a radio clock - or at least an interface that allows to select the appropriate time zone and switch daylight saving time on or off.

During the vehicle development process, the domains are organizationally represented by the design CoCs which design and release parts and components, and complete vehicle development which defines complete vehicle requirements and ensures that they are met during the PEP (compare Sect. 1.3.2). Figure 6.1 shows an excerpt from a detailed requirement catalogue that is used in practice to specify complete vehicles. Reference [2] provides extensive methodological support for the clarification of customer requirements.

Axiomatic Design is a methodological approach using elementary matrix methods that is used to attain stable system and sub-system concepts, and to analyze and optimize the transformation of customer needs into functional requirements, design parameters, and eventually into process variables [1]. Likewise, *Quality Function Deployment* (QFD) is a method that supports designers in transforming customer requirements into quality functions.

BMW Group	REQUIREMENTS CATALOGUE COMPLETE VEHICLE	Issue: 5
	Check List for Customer Relevant Vehicle Characteristics	Page 35

2   ACTIVE SAFETY / AGILITY / DYNAMICS
2.4   CHASSIS CHARACTERISTICS
2.4.2   CORNERING STABILITY

Specifications (what)	Evaluation	Standards (how)
2.4.2.1. Self-steer characteristic - constant radius o Characteristic function for assessing self-steering behaviour o Maximum stationary lateral acceleration o Assessment of dynamic driving behaviour: understeering, oversteering, neutral (Cross reference point 3.1.3.3)	SZ	EHB EG-4 No. 43, 56 EHB EF Chap 3.4.1 Steady-state cornering ISO 4138 or "pseudo- stationary" cornering ADS Sim. calc.
2.4.2.2. Cornering stability on constant radius - throttle inputs o Characteristic functions of vehicle load change sensitivity and course deviation o Yaw speed change caused by lateral acceleration o Cornering behaviour information	SZ	EHB EG-4 No. 43, 56 EHB EF Chap 3.4.1 VB EF-3 269/84 ADS Sim. calc. ISO 9816
2.4.2.3. Cornering stability - uneven road (e.g. individual obstacles or random unevenness) o Change in characteristic function of undisturbed cornering o Vehicle course deviation o Cornering behaviour information	SZ	EHB EG-4 No. 43, 56 EHB EF ADS Sim.calc.

Fig. 6.1 Excerpt from a complete vehicle requirements catalogue (Source: BMW)

6.1.2 Target Agreement

While vehicle requirements depict desirable (or undesirable) product properties or characteristics, vehicle targets describe the future status that should be accomplished (or avoided) by appropriate actions. The process of formulating, coordinating and pursuing a consistent set of product targets is referred to as *target management*. This also includes *target change management* as well as monitoring and reporting the level of target achievement. Target management starts during the early development phase and formally ends with the complete vehicle sign-off proving that all targets are met – all along the upper layer of the V-model shown in Fig. 1.6.

Conceptually, target management is the process of translating customer requirements from the language of marketing into the language of design. The requirement "acceptable knee space for rear passengers" e.g. can be converted into a formal target, consisting of an attribute (e.g. "L48 second row knee clearance") and its value (e.g. "9 mm") [3]. As over the PEP traceable decisions have to be made on the basis of product targets, these targets have to fulfill fixed requirements concerning form and content. To allow validation at any point in time, targets should be *hard targets*, meaning that they are quantitatively measurable (e.g. "braking distance on dry road from 100 to 0 km/h = 35 m"). If this is not possible, *soft targets* (e.g. "exterior vehicle styling is sportive and dynamic") must be evaluable by an agreed set of objective criteria. For the given example, these would be e.g. the ratios of wheelbase to wheel diameter, front and rear overhang to vehicle length and vehicle height to width (see Fig. 6.2). Vehicle targets have to respect multiple boundary conditions, such as general or specific laws, national standards and regulations, competitor's activities, constraints stemming from components shared by other variants, or restrictions due to the existing production environment.

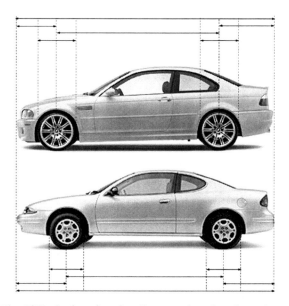

Fig. 6.2 Evaluation of a soft styling target by ratios of complete vehicle proportions

In order to make the vehicle defined by the product targets competitive, vehicle characteristics are compared with current competitors in a *strategic product characteristics profile* which evaluates all customer relevant characteristics of the competing vehicles in a reasonable level of detail. Then the profile of the newly planned vehicle is shaped by determining its level of differentiation from the core competitors. Each vehicle characteristic in the strategic product characteristics profile must be rated regarding its required level of differentiation on the market. There are three possible levels of differentiation:

- No/low differentiation ("acceptable standard"): For this characteristic, the new vehicle will offer the same standard as the basic competitors
- Medium differentiation ("among top 3"): This characteristic is not distinct for the brand but essential for the positioning of the new vehicle. It will be at the level of the best core competitors in the segment
- Strong differentiation ("best in class"): The characteristic is important for the brand profile and will be better than the core competitors

The individual targets deduced from this profile may initially conflict technically and/or financially (e.g. "trunk capacity = 400 l" vs. "L48 second row knee clearance = 9 mm"). Solving these conflicts by negotiating what marketing would like to have versus what the designers are able to provide is an important part of the target management process.

As the resulting set of targets has been derived from a comparison with existing competitor cars, it embodies a vehicle that would be competitive in today's market. To make sure that the new vehicle will still be attractive when it is launched years into the future among the successors of today's competitors, the targets are additionally evolved according to a forecast of what the standard will be at that time. The formal agreement of the product targets at the end of the target derivation process summarized in Fig. 6.3 is a pre-condition for passing the *target agreement* milestone described in Sect. 3.2.2.

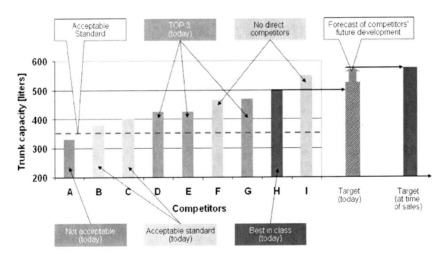

Fig. 6.3 Target derivation process (Source: BMW)

6.1.3 Sign-off Process

The maturity of the requested functions and properties of the complete vehicle related to target agreement is evaluated and proved by the sign-off process, the upper level process in the V-model (compare Fig. 1.6). During the sign-off process, the degree of complete vehicle target achievement is evaluated and then documented at defined milestones of the PEP, e.g.:

- Target frame confirmation
- Business plan confirmation
- Target agreement
- Concept confirmation
- Function confirmation
- Product confirmation
- Launch confirmation
- Process capability confirmation

For all agreed complete vehicle targets, the level of achievement is evaluated and quantified by use of a detailed conformity index (see Fig. 6.4).

	Evaluation result	Consequences
10	Outstanding	Totally exceeds expectations
9	Excellent	Slightly exceeds expectations
8	Very good	Fully meets expectations, delivering customer satisfaction (100%)
7	Good	Minor deviations from the nominal specification are detectable, customer is aware only in rare / isolated cases
6	Barely satisfactory	Irritating problems are apparent to the customer - failing to fully meet expectations
5	Unsatisfactory	Items will be repaired at the next opportunity
4	Deficient	Leading to customer disappointment, requiring an urgent visit to the workshop
3	Absolutely unsatisfactory	High customer annoyance
2	Breakdown	Extreme customer annoyance, legal or publicity action is threatened
1	Hazard, or vehicle ineligible for registration	Fails to meet safety requirements

Fig. 6.4 Complete vehicle target conformity evaluation scale (Source: BMW)

For early phase sign-off, it is not so much the actual deviation from the agreed targets but the comparison with the expected degree of deviation at the specific point in time in the PEP that denotes the maturity of the complete vehicle and indicates required corrective action.

6.2 Design Problem Management

If over the PEP tests show that vehicle functions or properties do not meet a set target, this non-conformance is reported as a formal *design problem*. From the series development phase on, problems are listed and addressed by *problem management*. Typically, problems are detected and reported by the complete vehicle integration processes (see Sect. 1.3.2.2): A typical example for a problem reported by geometric integration is the collision shown in the center picture of Fig. 4.6. Problems reported by production integration denote difficult or impossible production processes (compare Sect. 8.1) such as a bolt that can not be properly accessed with the intended power tool. The majority of problems however are reported by functional integration and denote functional targets which are not met by the product. Functional problems occur in each area of complete vehicle functionality (see Chap. 7) – most dramatically when the test results did not meet the legal targets, e.g. in a failed crash test.

When a problem is reported, it is fed into a problem management system and enters a formal problem elimination process. Here, the status of resolution for every problem is tracked via distinct progress levels, e.g.:

- Level 1: Problem is identified, defined and documented.
- Level 2: Responsible problem owner is identified; the problem evaluation process has been started.
- Level 3: Problem analysis is completed; root-cause has been identified.
- Level 4: Measures to avoid event recurrence have been defined and are confirmed; required part and/or process changes are approved and released.
- Level 5: Measures to avoid the problems are implemented; availability of changed parts and/or capability of changed process is confirmed.
- Level 6: Problem eliminated; solution effectiveness confirmed.

The primary mission of problem management is to make sure that every reported problem is completely eliminated as fast as possible or necessary. To do so, progress is checked regularly. For the most part, problems are normal elements of any development process and will be solved by the responsible designers as part of their routine. A common misconception is that criticality of a design problem always corresponds to its risk. But while high-risk problems usually get plenty of management attention and subsequently are solved as fast as possible, experience shows that problems of seemingly minor complexity can become very critical if not solved fast enough. This typically happens if responsibility for the problem has not been clarified or if the problem is not taken seriously ("wait for my new version" – see Sect. 4.2.1).

An efficient approach to detecting these time-critical problems is to compare their progress status with a generic timeline that represents a "normal" problem solution process. An example of such a timeline is illustrated in Fig. 6.5: The graph shows that a problem is expected to be reported the day of its occurrence (level 1) and have a problem owner responsible for the solution process named after four more workdays (level 2). The root-cause should be known 20 work days later (level 3) and the appropriate measures are expected to be confirmed after another 20 work days (level 4). Depending on the exact phase in which the problem occurred in (e.g. testing the first prototype build group), realization of the measures (level 5) can be expected 20 workdays after that. Depending on the availability of validation hardware, the planned overall lifetime of a problem from occurrence to proof of elimination (level 6) is about 15 weeks.

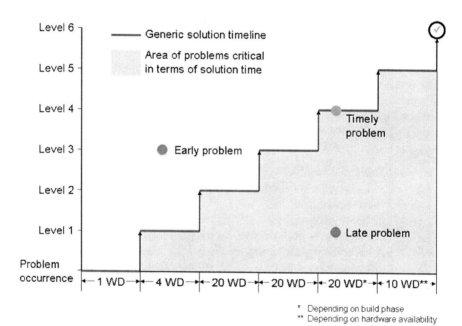

Fig. 6.5 Generic problem elimination timeline

To identify time-critical problems, the progress level of all reported problems is regularly checked against the generic timeline. Late problems (like the one indicated in the lower right corner of Fig. 6.5) are reported to management to initiate appropriate counteraction. A weak point of this process stems from the fact that it can only start after the problem has been reported and reached level 1. The actual day of occurrence is not known, so that a delay in reporting the problem can not be detected and escalated.

Together, the actual status and progress history of all problems in a vehicle project give a good picture of both the vehicle maturity and the effectiveness of the development process. This data is visualized as a problem landscape (see Fig. 6.6).

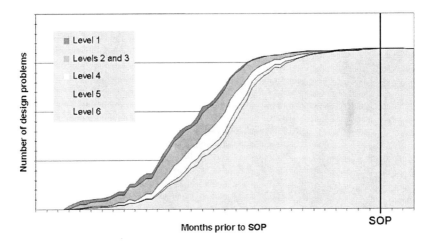

Fig. 6.6 Problem landscape

The problem landscape provides important information to management and allows top-level analysis such as e.g.:

- Were the problems reported during the first prototype build group eliminated in time or did they reoccur during the second one?
- Is problem solving generally faster or slower in this project than in comparable projects?
- Is one design CoC significantly slower in problem elimination than the others?
- Is the overall number of unsolved problems increasing or decreasing?

The last question is especially important during the second half of series development, when the product maturity should stabilize and tests are expected to confirm the state of the product rather than to bring up numerous additional problems. A practical index to evaluate this is the *relative problem solving speed*, which can be calculated as the ratio of the number of problems that have been solved during a specific period of time and the number of new problems reported during this period (see Fig. 6.7). This figure then can be periodically compared to a desired progression that has been derived from earlier projects.

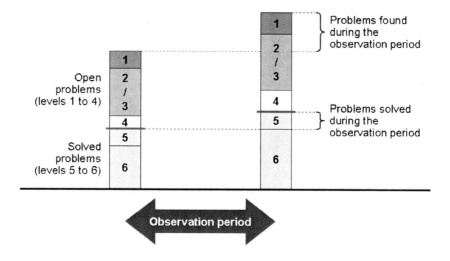

Fig. 6.7 Relative problem solving speed

6.3 Release and Change Management

6.3.1 Releases

The endpoint of any part design effort is a release. A release is an unambiguous description and documented publication of a component (including material, paint, leather, fabric, surface or software). The different release levels a part goes through all along the PEP are coupled to project milestones and indicate the part's maturity:

- During the concept phase, a *planning release* is the starting point for all planning activities at the beginning of a project. The planning release contains the expected series-production status of the part with reference to the target agreement. The geometry of the specific part is frozen and made public in the shared design workspace so that all designers involved in the project can utilize it.
- When the design is mature enough to build and test physical prototypes, a *test release* is the precondition to procure the required hardware and software. Test releases are issued separately for every prototype build group and can be valid exclusively for a certain prototype or a specific test procedure (e.g. a specific crash test).
- When a part's design has reached the required level of maturity for series production, a *preliminary release* is issued. A preliminary release is the precondition for ordering, designing and manufacturing series production tools. As lead

times for series tools vary greatly, preliminary releases are usually grouped in early, medium and late release groups.

- The final release level is the *production release*, the legally binding documentation of the designed part. Production release is based on the results of prototype and pre-series vehicle tests.

The designers in the COCs (compare Sect. 1.3.2.1) bear the personal responsibility and liability for the releases.

6.3.2 Design Changes

Design changes are what make engineering progress happen. Product development is a series of Design-Test-Evaluate-Change-loops on different levels. The reasons for changes and the impact they have on the overall product however vary greatly with the phase they occur in:

- During the concept phase, design changes are expected and welcome. This is the phase where designers lay the geometric and technical foundation for their solution and early changes usually improve the complete vehicle's conceptual quality. At this stage, a design change only means changing a 2D or 3D CAD model. Designers are expected to change their parts until a consistent vehicle concept is achieved.
- When – at the beginning of series development – the first hardware prototypes are built and tested, the gravity, more than the number, of the resulting changes is an indicator of the concept quality and maturity. Changes at this stage are expected to optimize parts and components rather than the complete vehicle concept. As prototype tools and parts must be changed, these changes require new test releases result in additional costs.
- While the first prototype build group is expected to highlight findings that will help to make the vehicle meet its targets, the second build group is only expected to prove that the car meets the targets. Time is running short for bigger design changes; as at this time a lot of the long lead-time series manufacturing tools already have begun to be manufactured and significant changes may require a new preliminary release, implying very costly adaptation of the series tools.
- During the launch phase, when the vehicle is already being produced in the plant as a pre-series car, changes should only include process related fine-tuning of individual parts. Any bigger change at this phase is a direct threat to the planned SOP date and implies high costs as tools and equipment may have to be changed.
- After SOP, design changes always bring with them a disturbance of the logistics and manufacturing system and hence represent a risk to product quality (compare Sect. 8.1). Intended changes during series production such as additional vehicle features and quality or cost improvements are closely evaluated both technically and financially before being released for series production.

Model year packages e.g. might include new options (especially new engines), that require comprehensive design changes and validation measures. Problems reported by quality management, customers or customer related tests (such as JD Power IQS) as well as results from continuous quality improvement lead to design changes that are – if possible[10] – collected and brought into series production as a change package once or twice a year to minimize validation and organizational effort. Cost improvement measures may be simple technical optimizations or material changes but can also include changing the supplier of a part – especially in international production, when a distant supplier is replaced by a local one to reduce production and/or logistics costs.

- Changes also may happen after *end of production* (EOP), when parts are only produced as spare parts for service and repair. As the volume drops significantly after EOP, most part manufactures reduce their production facilities which may make part changes necessary.

Depending on when over the course of the PEP it is required, changing e.g. the thickness of a body panel might mean changing an attribute, a 3D CAD model or a prototype part; it might also require repeating a crash test (which means an additional prototype), reworking a stamping tool, changing the parts supplier or even adding additional pieces of equipment to the body shop. The "rule of ten" [2] shows that the costs caused by a design change increases by factor 10 from phase to phase (see bottom line in Fig. 6.8) and makes it obvious that major design effort must be put into the concept phase to bring the product to maturity as early as possible and thus minimize the risk of late changes and the related cost risks. This so-called front-loading includes early use of simulation tools as well as methods for concept quality improvement such as *quality function deployment* (QFD) (compare Sect. 6.1.1) or *failure mode and effects analysis* (FMEA) (compare Sect. 7.9.2.2). Figure 6.8 shows the effects different design approaches have regarding the number and distribution of design changes over the PEP.

---

[10] If a problem however poses a direct safety threat to the driver or passengers, it requires immediate attention. Then, parts with the current design and all vehicles that are equipped with these parts are immediately blocked, vehicles which are already delivered are recalled. At the same time, a new solution has to be found, validated and implemented as fast as possible.

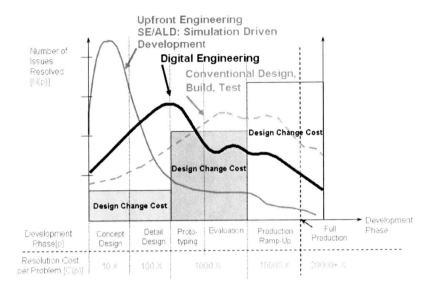

Fig. 6.8 Different frontloading approaches
(Source: International TechneGroup Incorporated [4])

6.3.3 Change Management

The interdependencies between the technical properties of all the parts, tools, equipment etc. that are needed in a vehicle project grow during each successive phase and make the product-production-system ever more complex. This complexity is the reason why, in a highly simultaneous engineering environment that includes product and process development, logistics and production throughout the whole sourcing chain, design changes must be applied for and comprehensively evaluated and communicated before being released and realized. *Engineering change management* is the process to identify, plan, control, document, communicate and decide changes throughout the overall vehicle lifecycle.

With its recommendation 4965 [5], the *German Car Makers Association* (VDA) has issued a comprehensive guideline for the change management of digital product data. Figure 6.9 shows phases and milestones of the recommended integral process.

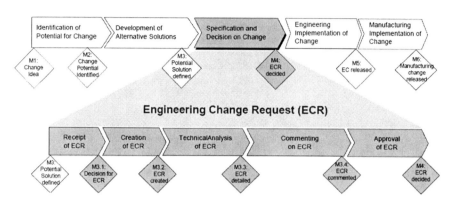

Fig. 6.9 Phases and milestones of an engineering change request (Source: VDA [5])

6.4 Quality Management

6.4.1 Definition of Quality

There are not many terms in product development which have been discussed for as long and as controversially as the term "quality". The following personal definitions by international quality experts may give a hint to the extent of the debate:

- Conformance to Requirements [6]
- Fitness for use [7]
- Products and services that meet or exceed customers' expectations [8]
- The loss a product imposes on society after it is shipped [9]
- The degree to which a specific product conforms to a design or specification [10]

Today, ISO 9000 features a generally accepted definition: Quality is "the totality of features and characteristics of a product or service that bear on its ability to satisfy stated and implied needs" [11]. The common denominator in all these definitions is the appreciation of the product by the customer. For the automotive industry, this appreciation can be divided in three main parts:

- *Concept quality*: The vehicle concept (including the price) should appeal to the target customer and offer enough reason (or emotion) to outpace its competition.

- *Production quality*: After parts manufacturing, assembly and shipping, the actual vehicle the customer receives should match the design specification and thus offer full functionality.
- *Long-term quality* or *reliability*: Even after years of service, the vehicle should maintain its functionality.

6.4.2 Pre-delivery (Internal) Quality Assessment

As shown in Sect. 3.2, the core job during the concept phase of a new vehicle or product line development is to anticipate the future customers' requirements, expectations and emotions and derive from it a consistent concept. At the end of the concept phase, the agreed complete vehicle targets must incorporate what management expects to be most appealing to the customer the day the car is offered on the market. As long as there is no "real" customer available to express his or her satisfaction with the car, the agreed set of targets defines what is considered 100% quality – an obviously very home-brewed definition of quality. But when during series development complete vehicle quality is assessed, the set of agreed targets is all that can be used for assessment. Bluntly spoken: All types of quality approval before the actual delivery mean nothing but that the assessed car meets what the manufacturer thinks are the customer's requirements.

During series development, functional integration ensures that the concept and its confirmed functionalities and features are realized by the final design. Production integration has to make sure that production transforms this design into real cars without any quality glitches. Complete vehicle testing then represents the customer. In development, the quality of a vehicle is measured by means of quality metrics that are computed in quality audits by adding up weighted quality defects. Depending on the extent of the audit, different quality figures can be defined. A first approach is static audits of the different vehicle areas:

- Body in white: Panel shape and gaps
- Outer surface: Paint and panel finish
- Interior and exterior trim: Trim fitment
- Engine bay: Fitment of components
- Under floor: Fitment of components

To include the quality defects that are only perceived when the car is in use, dynamic audits are carried through, e.g.:

- System test: Functional check of all electrical systems (see Sect. 5.2.6)
- Road test: Noise, vibration and harshness
- Road test: Dynamics, agility
- Water test: Water ingress

In all these different areas, trained auditors check every part of a vehicle against the required specification. If any differences occur, the auditors assess how serious that defect is using the conformity evaluation table shown in Fig. 6.4. The more serious a defect, the higher the quality figure.

As they measure the degree of compliance of a real car with the respective requirements, quality figures denote the maturity of the car design at a certain point in time. Out of the observation of different vehicle projects, a typical characteristic of quality figures over the course of the PEP can be derived: When the first prototypes are assessed, the quality figures are typically very high. Over later prototypes, pre-series cars and ramp-up production, product and process improvements then lead to a stepwise decrease until quality figures reach series production levels (see Fig. 6.10). In an ongoing vehicle project, these expected values can be compared with the actual quality figures to ensure a "normal" quality development [12].

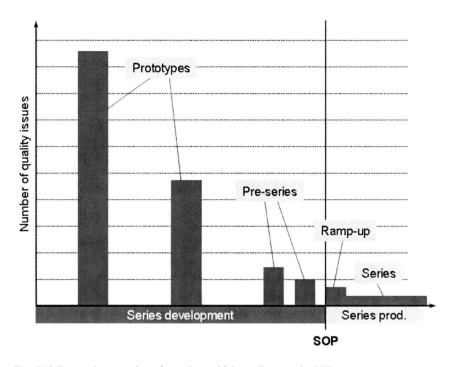

Fig. 6.10 Expected progression of complete vehicle quality over the PEP

During series production, quality figures taken from daily quality audits are the control variables within a plant's quality control loop. Vehicles are selected at random from the assembly line immediately before sales and subjected to a customer relevant audit against corporate quality specifications. Centralized specification of the quality targets and training of the auditors are required to maintain an equal level of quality over all production plants.

6.4.3 Post-delivery (External) Quality Assessment

As mentioned earlier: Only when the vehicle eventually has been handed over to the customer will the actual quality – in terms of fulfillment of customers' expectations – emerge. Right after SOP, when the car first hits the market, carmakers are eager to know how well their product is perceived by the customer. The three main channels through which customer feedback can be acquired are the dealership network, articles in car magazines or newspapers, and quality surveys.

6.4.3.1 Customer Complaint Processing

The immediate interface to the customer is the dealership where he or she purchased the car and brings it back for service or repair. The dealer gets direct feedback from the customer – positive and negative. Customer complaints are recorded at the dealerships and then collected and analyzed by central quality departments to create a comprehensive picture of a vehicle's quality.

An important advantage of this form of feedback – especially compared to the two others mentioned below – is that all information is kept within the organization and is not communicated to the public. A disadvantage however is that this information channel carries almost exclusively negative experiences with the product. Information concerning what the customer really liked about his vehicle is not collected and analyzed in a comparable way to his or her complaints.

The picture of vehicle quality derived by information collected from dealers however can be blurred. As it is not always 100% clear whether a problem reported by a customer is a real quality issue (or e.g. is caused by misuse or negligence) and warranty cases are paid for completely by the OEM, some dealerships tend to treat these cases as warranty cases more than others.

6.4.3.2 Car Magazines and Newspaper Articles

As part of the market launch, press events are held and journalists from car magazines[11] and newspapers' automotive supplements are given the opportunity to test the new vehicle. As most of these journalists are experts with many years of experience, their test results and the articles then written about the car usually con-

---

[11] Important car magazines are e.g. Car and Driver (US), Automobile (US), Motor Trend (US), Road and Track (US), Ward's Auto World Magazine (US), European Car (US), Auto Express (GB), Car Magazine (GB), Autocar (GB), CarMag (JP), AutoBILD (GER), Auto Motor und Sport (GER), AutoNEWS (GER).

tain important information for the manufacturer – especially as press cars are usually among the first vehicles that are delivered. Tendencies of the one or the other magazine or journalist in favor of, or against, a certain brand can be compensated if the whole spectrum of international publications is evaluated.

But quality feedback to the OEM is of course only a side effect of this information channel. Its primary goal is to bring as much public attention to the new car as possible and thus have a considerable impact on public perception of it. Probably the best known example of this effect is the article issued by the Swedish car magazine Teknikens Värld in 1997 that showed the brand new Mercedes A class flipped on its roof during a common lane change maneuver (later called the "elk test"). When the Teknikens Värld reported this problem, Daimler-Benz initially tried to reject the results. But as the case became known worldwide, Daimler-Benz had to improve the car's stability control system which led to millions of Euros of additional development cost and a three month delivery stop.

6.4.3.3 External Quality Surveys

External quality surveys are conducted by marketing research institutes or independent non-profit consumer organizations. These surveys differ in the observed content (e.g. whether concept quality is included or not) and the age at which the vehicle is assessed. Table 6.1 gives an overview of common vehicle surveys.

Table 6.1 Selection of external vehicle quality surveys

Name	Description	Content	Country	Age	Interval
NCBS	New car buyer survey	Sales, product, concept, reliability	Europe, Japan	3 months	Annually
IQS	Initial quality survey	Reliability	USA	3 months	2x per year
APEAL	Automotive performance execution and layout	Concept, design	USA	3 months	Annually
ECSS	European customer satisfaction survey	Service, product, reliability	Europe	2 years	Annually
QAS	Quality audit survey	Reliability	Europe	1–3 years	Annually
VDS	Vehicle dependability study	Reliability	USA	3 years	Annually
CSI	Customer satisfaction index	Service, product	USA, Japan	1–3 years	Annually
ADAC	ADAC breakdown statistics	Breakdowns	Germany	3–6 years	Annually
TÜV	Mandatory technical inspection	Operating safety	Germany	3–11 years	Annually

Important surveys for North America are the *Initial Quality Study* (IQS) and APEAL, both conducted annually by JD Power and Associates. While IQS provides in-depth diagnostic information on new-vehicle quality after 90 days of ownership structured by manufacturer, assembly plant, model, and platform [13], the APEAL study identifies the features that consumers find most appealing about their new vehicles. Owners evaluate more than 100 attributes in categories including ride and handling, engine and transmission, and comfort and convenience [14]. Figure 6.11 shows the APEAL nameplate ranking for 2008.

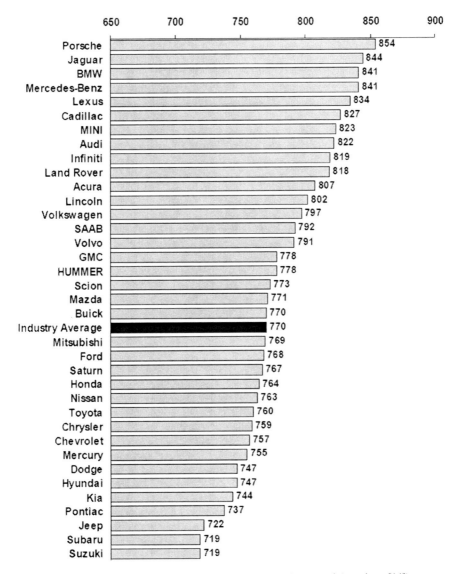

Fig. 6.11 2008 APEAL study nameplate ranking (Source: JD Power and Associates [14])

6.4.4 Quality Management Systems

A simple but practical definition of *Total Quality Management* (TQM[12]) is: "Everything you need to achieve world-class quality". To give a more detailed answer to the obvious question of what this would actually be, quality and standardization organizations have developed quality management systems – which are actually recommended sets of organizational structures, procedures and processes that enable an organization to identify, measure, control and improve its business processes and will ultimately lead to improved business performance.

The implementation of such a standardized quality management system is a huge effort and affects not only technical procedures but also parts of the company and management culture. Before corporate action can be changed, corporate thinking must be changed towards total quality management. This is why real management commitment is the indispensable prerequisite for implementing any quality management system.

To check whether and confirm that an implemented quality management system corresponds to the criteria of its developer (e.g. the ISO), it is audited regularly by an accredited auditor. The assessment is based on an extensive sample of its sites, functions, products, services and processes. If the quality system meets the requirements, the auditor will issue a certificate for each site that has been visited during the audit. QMS certificates must be renewed at regular intervals – usually around three years.

The most important fact to know about quality management systems and the according certificates is that they do not certify the quality of the product or the service rendered by the certified organization. At the end of the day, they only prove that the organization theoretically could deliver quality products.

6.4.4.1 The ISO 9000 Standard Family

The best known quality management systems are described in the ISO 9000 family,[13] of which ISO 9001 provides the appropriate requirements which an organization working in design and development needs to fulfill if it is to achieve

---

[12] ISO 9000 [11] defines Total Quality Management (TQM) as "a management approach of an organization, centered on quality, based on the participation of all its members and aiming at long-term success through customer satisfaction, and benefits to all members of the organization and to society."

[13] The ISO 9000 family consists of ISO 9000 (fundamentals and vocabulary. basics of quality management systems and core language of the ISO 9000 family), ISO 9001 (provides the requirements for organizations that design, develop, manufacture, install and/or service any product or provide any form of service), ISO 9002 (a subset of the ISO 9001 requirements for organizations that manufacture, install and/or service any product) and ISO 9003 (an even smaller subset of the ISO 9001 requirements covering solely final inspection).

customer satisfaction through products and services which meet customer expectations. Examples of these requirements are [15]:

- A consistent picture of all key processes in the organization including key points where each process requires monitoring and measurement
- Process control by monitoring, measurement and analysis to ensure that product quality objectives are met
- Quality objectives for each product the organization makes – which must be regularly measured and documented
- Clear requirements for purchased parts, components or services
- Methods to determine customer requirements and to communicate with customers concerning product information, inquiries, contracts, orders, complaints etc.
- Defined and documented stages of development with appropriate testing at each stage for every new product development
- Consistent performance reviews through internal audits and meetings to check whether the quality system is working and what improvements can be made
- Documented procedures for dealing with actual and potential, internal and external non-conformances
- A documented procedure to control quality documents within the organization that is known to every manager and associate in the organization

The actual quality system is documented in the quality manual, a formal statement from management that should be understood and followed by all employees at all levels. It contains measurable objectives for every employee to work towards. The quality system must be audited regularly to maintain and improve its effectiveness. Implementation of a quality system might include provision of suitable infrastructure, resources, information, equipment, measuring and monitoring devices, and environmental conditions.

The ISO 9001 process model is shown in Fig. 6.12. It represents an organization's quality management system in relation to customers' requirements and the achievement of customer satisfaction.

As the ISO 9000 family is a general standard, the proposed quality management systems are not tailored for the automotive industry. For this reason, ISO/TS 16949:2002 is an interpretation of ISO 9001 for the automotive industry agreed upon by major American and European car manufacturers. ISO/TS 16949:2002 contains the full text of ISO 9001:2000 and automotive industry-specific requirements [16].

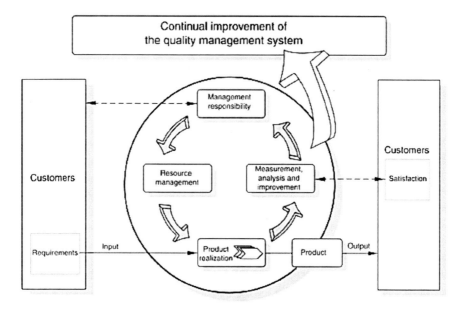

Fig. 6.12 ISO 9001 process model (Source: ISO [11])

6.4.4.2 Malcolm Baldrige National Quality Award

The Malcolm Baldrige National Quality Award is a national competition to identify and recognize U.S. companies excelling in quality and performance. It is
managed by the National Institute of Standards and Technology (NIST), a federal
agency within the Department of Commerce, and administered by the American
Society for Quality (ASQ). The actual award is traditionally presented by the
president of the United States.

Applications for the award include a comprehensive self-evaluation and are
evaluated by an independent board of examiners – composed of primarily private-
sector experts in quality and business. Examiners look for achievements and improvements in the following seven categories [17].

- Leadership – examines how senior executives guide the organization and how
  the organization addresses its responsibilities to the public and practices good
  citizenship.
- Strategic planning – examines how the organization sets strategic directions
  and how it determines key action plans.
- Customer and market focus examines how the organization determines requirements and expectations of customers and markets; builds relationships
  with customers; and acquires, satisfies, and retains customers.

- Measurement, analysis, and knowledge management – examines the management, effective use, analysis, and improvement of data and information to support key organization processes and the organization's performance management system.
- Workforce focus – examines how the organization enables its workforce to develop its full potential and how the workforce is aligned with the organization's objectives.
- Process management – examines aspects of how key production/delivery and support processes are designed, managed, and improved.
- Results – examines the organization's performance and improvement in its key business areas: customer satisfaction, financial and marketplace performance, human resources, supplier and partner performance, operational performance, and governance and social responsibility. The category also examines how the organization performs relative to competitors.

The criteria are designed to provide organizations with an integrated approach to organizational performance management that results in delivery of ever-improving value to customers and stakeholders, contributing to organizational sustainability, improvement of overall organizational effectiveness and capabilities as well as organizational and personal learning.

Organizations that pass an initial screening are visited by teams of examiners to verify information in the application and to clarify questions that come up during the review. Each applicant receives a written summary of strengths and areas for improvement [17].

Even though the award does not offer a standard model (as e.g. the ISO 9000 family), the criteria together with the actual evaluations implicitly describe a comprehensive quality management system. The successful implementation of this system is assessed by the evaluation process.

6.4.4.3 EFQM Excellence Model

As an equivalent to the Malcolm Baldrige National Quality Award in the U.S., the CEOs of some of the most prominent European companies founded the EFQM Excellence Model in 1989. Specifically designed to support all kinds of organizations in achieving business excellence, EFQM captures best practices world wide and documents them as a holistic yet practical model for the business community. Allowing individual approaches to sustainable excellence, the EFQM model provides a framework that allows companies to develop their vision and goals in a tangible and measurable way as well as to identify – and understand – their core business processes. In a self-assessment, companies can evaluate their current standing in terms of business effectiveness and efficiency. The fundamental concepts of the EFQM model are [18]:

- Results orientation: Excellence is achieving results that delight all the organization's stakeholders.
- Customer focus: Excellence is creating sustainable customer value.
- Leadership and constancy of purpose: Excellence is visionary and inspirational leadership, coupled with constancy of purpose.
- Management by processes and facts: Excellence is managing the organization through a set of interdependent and interrelated systems, processes and facts.
- People development and involvement: Excellence is maximizing the contribution of employees through their development and involvement.
- Continuous learning, innovation and improvement: Excellence is challenging the status quo and effecting change by utilizing learning to create innovation and improvement opportunities.
- Partnership development: Excellence is developing and maintaining value-adding partnerships.
- Corporate social responsibility: Excellence is exceeding the minimum regulatory framework in which the organization operates and strives to understand and respond to the expectations of their stakeholders in society.

The EFQM excellence model shown in Fig. 6.13 differentiates between enablers and results. While enablers cover what an organization does, results cover what the organization achieves – caused by the enablers.

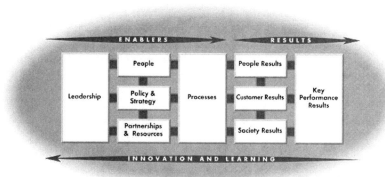

Fig. 6.13 EFQM Excellence model (Source: EFQM [18])

Today, EFQM is established as a powerful quality management system, and companies all over the world use its principles to bring their strategy and operations into alignment with their mission and goals.

Similar to the Malcolm Baldridge National Quality Award, the EFQM Excellence Award identifies and recognizes outstanding organizations in Europe and at the same time promotes their processes and structures as examples of excellence.

6.4.5 Quality Costs

Of course, optimum vehicle quality is desirable. But the big question for each carmaker is: At what cost? To give the correct answer, it is necessary to look at the overall picture of cost to obtain and cost caused by lack of quality. Essentially, the total cost of quality can be split into two fundamental areas: Costs of conformance and costs of non-conformance.

Costs of non-conformance – also called cost of poor quality – can be defined as costs that would disappear if products, processes and systems were perfect. Depending on where the non-conformance occurs, costs of non-conformance are divided into:

- Costs of internal failure – for failures that occurred at the carmaker's facility before delivery to the customer. Typical examples are scrap, rework, redesign, delays, shortages, failure analysis, re-testing, corrective action, waiting, down time, concessions, overtime, lack of flexibility etc.
- Costs of external failure – for failures that occurred after the car has been delivered to the customer, usually a result of not meeting the needs or specifications of the user. Typical examples are recalls, repair, complaints, warranties, returns, replacements, loss of customer's good will, losses due to sales reductions etc.

To achieve conformance, measures have to be taken. The affiliated costs are referred to as costs of conformance – or costs of good quality. Depending on when the quality measures are taken, costs of conformance are divided into:

- Costs of prevention – for pro-active measures to avoid defects including preventive maintenance. Typical examples are quality planning (prevention methods, reorganization of organizational structures and processes, product reviews), documentation (work instructions, technical instructions etc.), training (awareness, methods and tools, requirements and regulations), supplier evaluation, error proofing, capability evaluations etc.
- Costs of appraisal – for re-active measures to find defects by any kind of inspection. Typical examples are in-process and final inspection, checking and testing purchased parts and components, field tests, quality audits (internal and external), planning and inspection (cars, equipment, infrastructure, people), calibration of measuring equipment etc.

Experience shows, that appraisal and prevention are the key levers to reduce total quality costs. Armand V. Feigenbaum was among the first quality experts to argue that Total Quality Management ultimately results in the reduction of overall costs and induces a shift from failure and appraisal costs to prevention costs. The progression of the four cost elements over the four stages of TQM implementation is illustrated in Fig. 6.14.

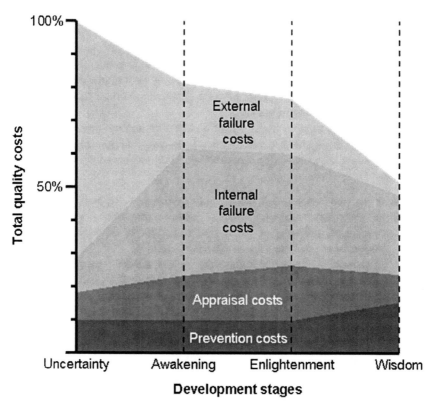

Fig. 6.14 Total quality cost progression over TQM implementation stages (Data Source: Educe [19])

Today, external failure costs are the biggest problem for carmakers. All OEMs struggle to reduce the costs for warranty and goodwill. In order to undertake effective measures upfront and reduce customer problems in the long run, meticulous tracing and reporting of warranty and goodwill cases, as well as appropriate methods to analyze these data are required. BMW Group e.g. uses a tool that allows graphical illustration of warranty and goodwill cases per vehicle. A set of curves indicates the number of cases for different vehicle ages compared to the date of production. A special algorithm then generates a forecast of the warranty and goodwill situation. In the sample graph in Fig. 6.15, the red line represents the number of warranty and goodwill cases of 24 month old vehicles, the blue line those of 12 month old vehicles, and the green one that of 6 month old vehicles – each over the respective date of production.

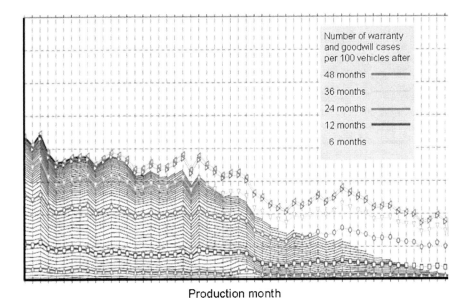

Production month

Fig. 6.15 Warranty and goodwill graph (Source: BMW)

References

1. Albano LD, Suh NP (1994) Axiomatic design and concurrent engineering: Tools for product introduction. Computer-Aided Design 26:499–504
2. Ehrlenspiel K (1994) Integrierte Produktentwicklung. Methoden für Prozeßorganisation, Produkterstellung und Konstruktion. Hanser, Munich
3. Grabner J, Nothhaft R (2006) Konstruieren von Pkw-Karosserien. Springer, Berlin
4. Lemon J (2008) Why simulation should drive product development.
    http://www.iti-oh.com/Education/Articles/SimLedDev.htm.
    Accessed 20 November 2008
5. VDA 4965 (2005) Engineering change management (ECM)
6. Crosby PB (1979) Quality is free: The art of making quality certain.
    New American Library, New York
7. Juran JM (1974) The quality control handbook, 3rd edn. McGraw-Hill, New York
8. Kano N, Seraku N, Takahashi F, Tsuji S (1984) Attractive quality and must-be quality. Journal of the Japanese Society for Quality Control 14(2):39–48
9. Taguchi G (1986) Introduction to quality engineering: Designing quality into products and processes. Quality Resources, New York
10. Feigenbaum AV (1983) *Total quality control*: 3rd edn.
    McGraw Hill, New York
11. ISO 9000 (2005) Quality management systems. Fundamentals and vocabulary
12. Weber J (1999) Optimierung des Serienanlaufs in der Automobilproduktion.
    VDI-Z 141:23–25

13. JD Power (2008) 2008 initial quality study results.
    http://www.jdpower.com/autos/articles/2008-Initial-Quality-Study-Results.
    Accessed 20 November 2008
14. JD Power (2008) 2008 APEAL study results.
    http://www.jdpower.com/autos/articles/2008-APEAL-Study-Results.
    Accessed 20 November 2008
15. ISO 9001 (2000) Quality management systems. Requirements
16. ISO/TS 16949 (2002) Quality management systems. Particular requirements for the application of ISO 9001:2000 for automotive production and relevant service part organizations
17. Baldridge National Quality Program (2008) Criteria for performance excellence.
    http://www.quality.nist.gov/Business_Criteria.htm. Accessed 20 November 2008
18. EFQM (2008) The fundamental concepts of excellence.
    http://www.efqm.org/uploads/fundamental%20concepts%20English.pdf.
    Accessed 20 November 2008
19. Wilson D (1995) The cost of quality.
    http://www.educesoft.com/quality/costofquality.htm. Accessed 20 November 2008

Chapter 7
Primary or Customer Relevant Complete Vehicle Characteristics

Abstract What counts in the automotive industry at the end of the day is whether a vehicle suits the needs or expectations of the potential customers. Characteristics such as costs, design appeal, cabin comfort, infotainment functionality, agility, passive safety, theft deterrence, reliability or sustainability are the main factors in the purchasing decision. But in their longing for technical perfection of singular components, designers all too often lose sight of what the customer really expects from his or her vehicle. Knowledge of the legal and personal requirements and their interdependencies, the design approaches to meet them and the methods to validate them is a prerequisite for successful automotive design.

Engineers often forget the most essential question of what vehicle characteristics the customer actually appreciates – and that eventually drive his purchase decision – and instead focus on the question of how these features can be achieved. At the end of the day, it is e.g. not the engine's power and torque, the aerodynamic drag or the total mass that the customer cares about; it is the vehicle's acceleration and maximum speed, the wind noise and the energy costs. According to the principles of axiomatic design (see Sect. 6.1.1), complete vehicle characteristics can be divided into three levels: The first level, the *customer domain*, is represented by the customer relevant or primary vehicle characteristics which directly influence the purchase decision. The functional or secondary vehicle characteristics on the second level, the *functional domain*, are the vehicle functions and properties that are necessary to realize the first level characteristics. The characteristics of the *physical domain* are physical prerequisites for the characteristics of the functional domain [1].

Customers however are e.g. typically not directly interested in the engine's performance or fuel consumption. What they actually care for (and what they mean when they ask the dealer about performance and fuel economy) is the vehicle's acceleration and operating costs. Figure 7.1 depicts the 10 vehicle characteristics that are directly perceived by the customer – and on which automotive development therefore should focus.

J. Weber, *Automotive Development Processes*, DOI 10.1007/978-3-642-01253-2_7,
© Springer-Verlag Berlin Heidelberg 2009

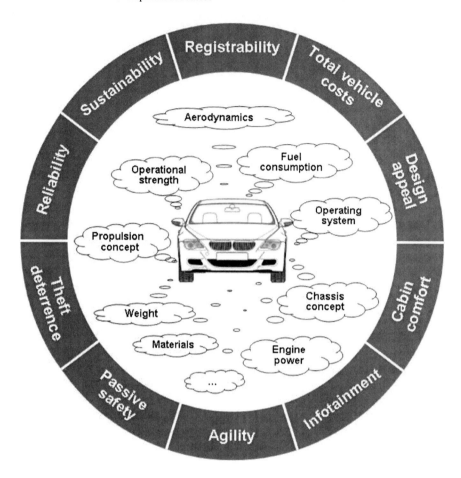

Fig. 7.1 Directly customer relevant complete vehicle characteristics

In order to meet the future customers' requirements, these customer relevant characteristics must be followed and analyzed according to the following structure:

- Legal and customer requirements: What are the actual – regulatory and individual – customer requirements? What are the disciplines in which the vehicle vies with its competition for their customers' favor? What are the properties and functions for which a customer would pay a certain premium?
- Component and system design: Which system architecture and component designs are the most suitable to optimally meet these customer requirements? Which design processes, tools and methods are most appropriate?
- System integration and validation: How can the behavior of a sub-system or the complete vehicle be tested and validated with regards to the customer relevant characteristic?

In contrast to most books on automotive development which focus on and are structured by the main vehicle components, this chapter – as the central part of this book – analyses in-depth and one-by-one each of the identified customer relevant vehicle characteristics over the following 10 sections. This approach dictates that some components are discussed several times, each time for their contribution to another complete vehicle characteristic. Taking tires as an example, Table 7.1 shows how one component's parameters can influence all complete vehicle characteristics.

Table 7.1 Tire's influence on complete vehicle characteristics

Complete vehicle characteristic	**Section**	**Contributing tire properties**
Registrability	7.1	Conformity with legal requirements
Total costs	7.2	Cost, lifetime, mass, rolling resistance
Design appeal	7.3	Dimension, color, sidewall design, profile design
Cabin comfort	7.4	Damping characteristics, rolling-off noise
Infotainment functionality	7.5	Integrated tire pressure sensors
Agility	7.6	Structure, dimensions, material, profile
Passive safety	7.7	Radial stiffness
Theft deterrence	7.8	Unequivocal identifier e.g. on sidewall
Reliability	7.9	Intrusion resistance, durability
Sustainability	7.10	Lifetime, recyclability of materials, sustainability of manufacturing and logistics

The tire example makes it obvious that component design is always a compromise between the various and often conflicting complete vehicle requirements. Understanding and optimally selecting among the choices implied by these compromises is the main work of development engineers. Technologies that allow better compromises give the respective component manufacturer a competitive advantage.

7.1 Registrability

7.1.1 Legal and Customer Requirements

The initial customer requirement concerning registrability is very obvious: Obtainment of registration approval for his or her vehicle in the respective market. For any vehicle manufacturer, it is as simple as this: No registration approval, no sales. Beyond this, conformance with all future regulatory requirements throughout the vehicle's lifetime can be seen as part its reliability (compare Sect. 7.9).

7.1.1.1 Development of International Vehicle Regulations

National regulations concerning on-road vehicles have developed since the beginning of the last century.[14] In 1909, Germany's *Law Concerning Road Traffic* came into effect, including detailed technical requirements concerning the availability and characteristics of components such as brakes, steering, lights, reverse gear, horn and a device to prevent unauthorized usage. This law was adopted in 1937 by the German StVZO, which was the basis for registration approval in Germany until 1995.

Initially, national regulations developed independently – especially in the two biggest markets, namely Europe and North America. With few exceptions, both markets sold local products and mutual alignment of regulations was unnecessary. Later, vehicle registration regulations not only reflected different ideas of how to achieve safety or eco-friendliness; they became also a practical means of import control.

In 1956, the *Working Party on the Construction of Vehicles* (WP.29), a subsidiary body of the *Inland Transport Committee* (ITC) of the UNECE concluded the so called *Rome Agreement* as a first step to not only dealing with the safety concerns posed by road traffic, but also with the hindrances diverse state regulations impose to the free flow of commerce across state borders.

This led to the establishment of the *UNECE 1958 Agreement* or *Geneva Agreement*, a legal framework[15] aimed at the removal of international trade barriers

---

[14] The first official regulation was the German "Law Concerning Road Traffic" which came into effect of May 3rd 1909. It included detailed requirements concerning the availability and characteristics of components such as brakes, steering, lights, reverse gear, horn and a device to prevent unauthorized usage. This law was adopted in 1937 by the German road traffic licensing regulations (StVZO), which is still the basis for registration approval today.

[15] The formal title of the document is: Agreement concerning the adoption of uniform technical prescriptions for wheeled vehicles, equipment and parts which can be fitted and/or be used on wheeled vehicles and the conditions for reciprocal recognition of approvals granted on the basis of these prescriptions.

within the automotive industry and which was ratified by 33 UNECE member states, as well as Japan, Australia, South Africa and New Zealand. It embraces (for vehicle systems, parts and equipment; not for the entire vehicle) 114 unified technical UNECE regulations, a unified type approval process (see Sect. 7.1.3), and reciprocal recognition of type approvals among contracting parties [2].

In the U.S., the *National Highway Traffic Safety Administration* (NHTSA) has a legislative mandate to issue safety standards and regulations – the so-called *Federal Motor Vehicle Safety Standards* (FMVSS) – to which manufacturers of motor vehicles and equipment items must conform and certify compliance. The FMVSS requirements are specified to reduce the likelihood of car accidents as well as the risk of death or injury in the event of an accident. The first standard to become active was the FMVSS 209 seat belt specification on March 1, 1967. Since then, over 85 standards have been established, divided in crash avoidance, crash worthiness, post-crash and other standards [3].

The *Motor Vehicle Air Pollution Control Act* is a 1965 amendment to the *Clean Air Act* of 1963. It set the vehicle emissions for the 1968 models as 72% less hydrocarbons, 56% less carbon monoxide and 100% less crankcase hydrocarbons compared to the 1963 levels.

Legal requirements are discussed in detail for each complete vehicle characteristic in Sects. 7.2–7.10.

The most notable non-signatories of the 1958 Geneva Agreement are the United States and Canada, adhering to the requirements stated in the FMVSS or the broadly similar CMVSS respectively. And still today, the U.S. and Canada do not recognize UNECE approvals – and vice versa. This means that without appropriate technical modifications, vehicles cannot be imported or exported between the U.S. and most of the rest of the world.

A typical example of such different requirements is headlights. In the U.S., compliance with FMVSS 108 is mandatory, while virtually the rest of the world requires headlights according to UNECE 37 standards. The main differences between the two standards are in the amount of glare permitted towards other drivers on low beam, the minimum amount of light required to be directed straight down the road, and the specific locations within the beam at which minimum and maximum light levels are specified. As each system's advocates claim the other one unsafe, vehicle owners who want to ship their U.S. car to Europe or vice versa have to change headlights to get their car registered. If headlights according to the required specification are not available, a car can not be registered at all.

In an attempt to improve this situation, the *UNECE 1998 Global Agreement* was negotiated and concluded under the auspices of WP.29 and the leadership of the European Community, Japan and the United States of America in 1998. Even though it does not claim reciprocal recognition of type approval, the Agreement establishes a process through which the participating states can jointly develop *Global Technical Regulations* (GTRs) addressing safety, environmental (air and noise pollution emission), energy and anti theft requirements for components such

as exhaust systems, tires, engines, acoustic shields, anti-theft alarms, warning devices, child restraint systems etc. In March 2000, WP.29 became the *World Forum
for Harmonization of Vehicle Regulations*. Despite all efforts to harmonize national regulations, the total number of motorized vehicle regulations has dramatically increased on a global basis. Figure 7.2 depicts the historic and expected progression.

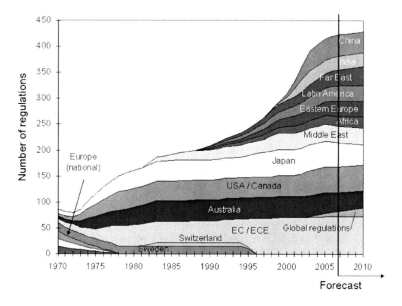

Fig. 7.2 Increase in international motor vehicle regulations (Source: BMW)

7.1.1.2 Specification of Technical Requirements

In most countries, regulatory requirements for registration approval apply to
technical specifications within the following areas [4]:

- Active safety (accident prevention)
  - Brake systems
  - Driver's field of vision
  - Tires
  - Lighting
  - Electro-magnetic immunity
- Passive safety (injury mitigation)
- Emissions
  - Exhaust gas
  - Noise

- – Evaporation
- – Electro-magnetic
- Theft deterrence
- Position of license plate, manufacturer's plate, VIN etc.
- Engine power
- Weights and dimensions

Regulatory requirements may concern the customer domain (e.g. passive safety), the functional domain (e.g. emissions), and the physical domain (e.g. vehicle dimensions) – compare Fig. 7.1). The actual requirements in each of the categories shown are listed in the following sections.

7.1.2 Component and System Design

Designing street-legality into a vehicle means meeting the legal requirements. For automotive OEMs and suppliers, the complexity of this homologation process has three dimensions:

- Legal requirements and the corresponding approval process are different in different markets.
- Legal requirements are a cross section through all kinds of vehicle characteristics such as mass, dimensions, emissions, passive safety, acoustics etc.
- Legal requirements are changing fast.

Management thus not only has to be aware of – and comply with – the total of all the vehicle regulations in all targeted markets at the time of start of production of each vehicle, but also to closely follow the future development of these regulations to make sure that the vehicle in question can be registered throughout its lifetime. If they fail to do so, car manufacturers risk confronting their customers with additional costs of ownership (e.g. fees or additional taxes for non-compliance with changed emission limits), considerable impairment of use-value (e.g. limited access to city centers due to newly implemented restrictions for cars exceeding certain dimensional limits), a drastic drop in the vehicle's resale or trade-in value, or – at worst – loss of street-legality (e.g. through non-conformance with elevated safety standards).

Some regulations (especially those concerning vehicle emissions) highly depend on national governments' current policies and are hence relatively volatile. For this reason, representatives of the automotive industry on national level like the *German Car Makers Association* (VDA), the *American Automobile Manufacturers Association* (AAMA), or the Japanese Automobile Manufacturers Association (JAMA) and supranational level like the *European Automobile Manufacturers Association* (ACEA), or the *International Organization of Motor Vehicle Manufacturers* (OICA) – exert political influence on the respective governments in order to jointly create a stable basis of responsible, realistic and reasonable legal requirements.

To cope with the multitude of regulations, the task of ensuring compliance is usually performed by specialists in different roles: *Project leaders for homologation* are responsible for homologation within a specific vehicle project, and coordinate registration approval issues for this project from the beginning of development until the start of production. *Homologation specialists* are knowledgeable in international legislation for one special field of regulations (e.g. emissions). Within this area, they are the main contact point and keep their partners up-to-date concerning worldwide legislation trends and their impacts on product development. They perform system approvals in coordination with the project leader for homologation and the respective *national specialists*. The expertise of national specialists lies in their specific knowledge of the legal requirements of one or more countries. They are the main contact point especially for homologation partners at sales subsidiaries and importers, coordinate approvals of all products for one country, and keep their partners up-to-date concerning legislative trends in the specific country.

7.1.3 System Integration and Validation

Type approvals are governmental means to make sure motorized vehicles are safe to use on the road and respect environmental considerations, without having to in-spect and test every single one. Just like the technical requirements, the necessary processes to obtain type approval for motorized vehicles also generally differ be-tween Europe and North America.

7.1.3.1 Europe

In Europe, two parallel type approval schemes have exist for over 20 years:

- *EU Motor Vehicle Type Approval* according to *EC Whole Vehicle Type Approval* (ECWVTA), EU directive 2007/46/EC as part of framework directive 70/156/EEC for type approval of whole vehicles
- *UNECE Type Approval* according to UNECE regulations for type approval of vehicle systems and separate components (but not whole vehicles)

Both schemes require applicants to formally apply to an accredited approval authority. Prior to testing, the applicant is required to submit technical documentation according to Annex I or III that adequately specifies the product in terms of the relevant features. These documents act as a check list for the sample of the product that is provided for testing and – after the test – as the basis for the approval itself.

To establish compliance with the corresponding EU directive or UNECE regulation, a prototype of the vehicle is tested by a recognized technical service which must be accredited to EN 45001 and be appointed by the national approval authority to test on its behalf. The technical service issues a test report according to Annex V which has to be submitted to the approval authority by the applicant. Also, both schemes require conformity of production according to Annex X. The approval authority must verify that adequate arrangements for ensuring conformity of production have been taken by the applicant. This is achieved by submission of a certificate according to ISO 9001, ISO 9002 (compare Sect. 6.4.4.1) or an equivalent standard to the authority.

If both the vehicle test and the QMS assessment are passed, the approval authority issues an *EC Type Approval* document according to Annex VI, a list of test results according to Annex VIII, and a *certificate of conformity* according to Annex IX.

7.1.3.2 U.S. and Canada

In the U.S. and Canada, it is the sole responsibility of the manufacturer to certify that the vehicle is in full compliance with the relevant FMVSS or CMVSS standards. As opposed to the European type approval, this is a self-certification process. In order to provide certification, the manufacturer must take all appropriate actions, such as laboratory testing in accordance with the FMVSS or CMVSS or conducting other studies or analyses to ensure that the vehicle fully complies. NHTSA or Transport Canada respectively review these certifications with series cars and parts.

In analogy to the European process, production conformance has also to be assured. To ensure continued compliance of vehicles throughout the production run, the vehicle manufacturer has to establish an effective quality control program and periodically inspect and test vehicles randomly selected from the assembly line to ensure that the original performance is carried through to all other units. Detailed realization of the quality program is not specified by NHTSA or Transport Canada but left to the manufacturer [5].

In addition to this safety-related approval, vehicle manufacturers need to prove compliance with the relevant regulatory agencies, e.g.:

- U.S. Environmental Protection Agency (EPA)
- California Air Resources Board (CARB)
- Northeast Trading Region (NTR)
- Environment Canada

Manufacturers have to undertake an exactly defined and controlled emissions type approval in every model year. The durability of the components must be proven. This procedure takes approximately 18 months. The fuel consumption which was determined in the emissions approval is re-checked by the regulatory agency.

7.2 Total Vehicle Costs

7.2.1 Legal and Customer Requirements

Customer requirements concerning total vehicle costs are easy to explain: For a given product, the costs associated with buying, owning and operating the car should be as low as possible. In most cases however, customers base their purchase decision mainly on the purchase price – and pay only minor attention to the costs that occur later while the vehicle is being used. But even for a brand new vehicle, owning it can be actually more expensive than buying it [6]. If advertised clearly, total vehicle costs are an essential vehicle characteristic for a potential customer's purchase decision.

In general (e.g. [7], vehicle costs are broken down into two main categories: *Fixed costs* and *variable costs*, with fixed costs being independent and variable costs being dependent on the actual usage of the car (where usage mostly is a synonym for mileage). From a customer's point of view, it is also important to differentiate vehicle costs by the point in time in the life cycle at which they occur: At the beginning (that is in connection with the acquisition of the vehicle, during the actual period of ownership and utilization or at the end that is when the car is resold or decommissioned). Considering this, there are four main categories of vehicle costs which must be differentiated and will be discussed in the following sub-subsections: Acquisition costs, costs of ownership, operating costs and end-of-ownership costs.

7.2.1.1 Fixed Costs (Costs of Ownership)

Acquisition Costs

Acquisition costs are the total one-time investment the customer has to make in order to become the owner of a car. This includes:

- Invoice price incl. taxes, discounts, delivery charges, custom duties etc. (when vehicle is purchased directly)
- Down payment incl. taxes, discounts, delivery charges, custom duties etc. (when vehicle is financed)
- Registration fees
- Deductibles from taxable income (if applicable)

Running Costs of Ownership (Zero Mileage Costs)

Monthly or annual fixed costs that are independent from mileage or wear and occur only for having a car ready to be used:

- Vehicle taxes
- Insurance (liability, collision, comprehensive, uninsured/underinsured coverage, loss of use, loan/lease payoff, etc.)
- Automobile club / roadside assistance membership (e.g. AAA, ADAC, RAC)
- Periodic vehicle safety and emissions inspections (if required)
- Residential parking
- Interest rates (when vehicle is financed)
- Lease rates (when vehicle is leased)

End-of-ownership Costs

The costs (and revenues) that occur when selling or disposing of the car.

- Resale value (when vehicle is purchased)
- Costs of reselling (when vehicle is purchased)
- Wrecking costs (if applicable)
- Mileage allowance exceedance fee (if applicable when vehicle is leased)

7.2.1.2 Variable Costs (Operating Costs)

Variable costs that depend on the extent to which the car is actually used:

- Fuel / energy
- Coolants, lubricants and other consumables
- Maintenance / renewal of wear parts (tires, spark plugs, wiper blades etc).
- Washing and cleaning
- Parking
- Street tolls, zone related traffic fees (e.g. London Congestion Charge, Singapore Area Licensing Scheme, New York Congestion Pricing)
- Fees for ferries or auto-trains
- Repair
- Tickets and fines

7.2.2 Component and System Design

While reduction of manufacturing or development costs is an established process at every OEM and supplier, there is usually no systematic approach to reduce the total vehicle costs incurred by the customer. In order to reduce total vehicle costs for the customer, development has to identify the levers within each of the cost categories discussed above.

7.2.2.1 Reduction of Costs of Ownership

Reduction of Acquisition Costs

Acquisition costs – and the resulting interest or lease rates – can mainly be influenced by the actual manufacturing costs and conformance with requirements:

- Lower production costs through design for cost and design for manufacturing result in lower manufacturing costs and thus allow lower sales prices.
- Compliance with requirements resulting in the reduction of taxes or fees imposed on the manufacturer, e.g. local content requirements to avoid import duties, fuel consumption limits to avoid *corporate average fuel economy* (CAFE) fees (compare Sect. 7.10.2.1), compliance with technical requirements necessary to obtain tax reductions).
- Manufacturing the vehicle in the same region where it is to be sold saves transport costs for the complete vehicle.

Reduction of Zero Mileage Costs

For financed and leased cars, the largest portion of the costs of ownership is interest rates or lease rates respectively. Both depend mainly on the acquisition costs of the vehicle and can be lowered with the levers mentioned above. Taxes and insurance fees (both usually included in lease rates) can be reduced by the following measures:

- Reduction of vehicle tax by keeping the relevant technical parameters (e.g. engine volume, weight, emissions, overall dimensions etc.) within the critical limits and implementing technology that makes the vehicle eligible for tax reductions (e.g. hybrid engines, compare Sect. 7.10.2.3).
- One of the main factors influencing a vehicle's insurance cost is a risk rating deduced from statistically calculated damage likelihood. It depends on two parameters: The level of active safety of the car (determined by vehicle design, agility, cabin comfort, and infotainment functionality as the four vehicle characteristics that make a car safe to drive and thus reduce the likelihood of getting involved in an accident) and the average driving performance of its typical owner. While the first can be influenced through technical features (see the respective sections in Chap. 7), the only lever for the latter parameter is to give the car a design that is less appealing to high-risk drivers.
- A parameter that particularly drives full coverage insurance fees is a relative repair cost rating provided by insurance companies. This rating depends on

how costly repair for typical vehicle damages is. Repair costs can be lowered by a car design that allows easy damage analysis, fast exchange of damaged parts and avoids costly highly integrated parts that must be fully exchanged when partially damaged (compare Sect. 8.2).
- Fees for theft insurance can be influenced by the level of attractiveness the car has for thieves and by the theft-deterrence features it is equipped with (see Sect. 7.8).
- Including roadside assistance and other services with the purchase a new vehicle avoids the necessity of third party auto club membership and the related fees.
- Particularly in urban areas, costs for residential parking can depend on the outer dimensions of a vehicle.

Reduction of End-of-ownership Costs

For a newly purchased car, the most difficult to calculate – and underestimated – cost is the depreciation, or the difference between invoice price and the resale value at the end of the ownership period. Not having to care about reselling the car at the end and thus having cost transparency from the beginning is one of the aspects that make car leasing so attractive.

Neglecting physical damages, the main parameters that influence the resale value of a car are its mileage, brand value, general condition and reliability. As mileage is independent from a vehicle's design, characteristics that increase the resale value and thus leverage the depreciation are:

- A timeless styling that remains appealing for a long time
- Conformance to forecast future legal requirements
- An open system architecture that allows integration of new features / technologies (e.g. new infotainment systems)
- Design for reliability: A vehicle that is well-known for its longevity keeps a high resale value (compare Sect. 7.9).
- A high brand value (compare Sect. 2.2)

7.2.2.2 Reduction of Operating Costs

Apart from the ever increasing costs for energy, operating costs of a vehicle can be influenced by a variety of levers:

- Provision of a fuel-efficient design (see Sect. 7.10.2.2)
- Reduced need for exterior washing through aero-dynamic design, dirt repelling surfaces and dirt-insensitive colors
- Reduced effort for interior cleaning through dirt repelling fabrics and textures

- Reduction of maintenance and costs through design for service and usage of industry-wide standardized wear parts
- Costs for speeding or parking violations can be reduced through assistance systems (compare Sect. 7.5) that remind the driver when driving over the speed limit or help him to find a (legal) parking lot
- Consideration of dimensional, weight and emission limits that trigger road tolls and fees for auto trains, ferries etc.

7.2.3 System Integration and Validation

While there are plenty of methods to evaluate distinct cost types such as fuels or insurance, there is no standardized approach to rate the overall cost efficiency of a vehicle. The unknown cost associated with depreciation or repair costs seemingly make this effort senseless. But with total costs being a major aspect in a purchasing decision, standardized vehicle utilization scenarios – equal to the driving cycles used for emission and fuel consumption testing which are discussed in Sect. 7.10.2.5 – could be a basis for meaningful total vehicle cost ratings that allow customers to realistically compare the total financial burden imposed by different vehicles.

An approach in this direction is the vehicle cost table *Your Driving Costs* that is published and yearly updated by AAA since 1950 [8]. For five different vehicle types,[16] it estimates both yearly costs of ownership and operating costs per mile – based on average costs for fuel, maintenance, tires, insurance, licensing, registration, taxes, depreciation, and financing. Figure 7.3 depicts the resulting total yearly costs over the mileage.

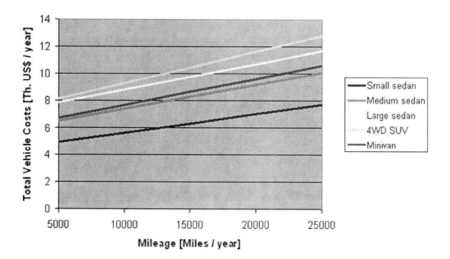

Fig. 7.3 Total vehicle costs (Data source: AAA [8])

---

[16] Small sedan, medium sedan, large sedan, SUV, minivan.

7.3 Design Appeal

7.3.1 Legal and Customer Requirements

Perception of a new car by a potential customer usually happens from the outside to the inside and along different levels of detail: The first look catches the vehicle's body style and proportions. Is it a classic roadster? A dynamic coupe? A luxury sedan? As the customer gets closer to the car, surfaces come into focus. The shape of the hood, characteristic swage lines in outer body panels, the door. Eventually, details such as door handles, front- and rear lights or exterior trim parts ere experienced. Opening the trunk lid gives a first impression of practicality and eventually entering the car allows the potential customer a first experience of ergonomics by means of the ease of access.

Then, from inside, the vehicle's interior can be explored – again starting with proportions: Does the interior feel spacious or restricted? Is the dashboard a piece of modern art furniture or rather a rack for vehicle equipment? Are the seats sofas rather than stools? Next, interior surfaces like trim parts or upholstery give an impression of value. The details of interior design such as control elements, displays or package trays allow further experience of ergonomics and value perceived. Figure 7.4 illustrates this process of experiencing a vehicle's design.

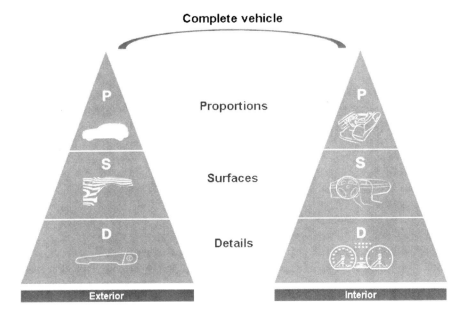

Fig. 7.4 Levels of vehicle design perception (Source: BMW)

Out of this process of perception, three main customer relevant properties can be deduced which are determined by the vehicle's conceptual design: Styling, ergonomics and practicality.

7.3.1.1 Styling

As discussed above: The first characteristic of a car that should catch a potential customer's attention, ignite his emotional perception and eventually draw him to the showroom is the aesthetic appearance of its exterior styling [9]. Typical features that determine a vehicle's exterior styling are: Overall dimensions, wheel base, wheel size, track, front and rear overhang, hood line, windscreen line, windows, pillars, belt line, doors, wheel house size, color etc. Brand specific styling features differentiate the vehicle from its competitors.

Both from the customer's and society's viewpoint, styling carries an important statement about a vehicle's owner's character. And even if this statement is understatement: For most customers, the message sent out by their vehicle's styling is as important as its objective technical features (such as performance or acoustics etc.). This message can be rather weak or strong. Figure 7.5 shows four cars with a distinctive exterior styling that send very strong (though very different) messages to its environment, e.g. provocation and radicality by a Lamborghini Reventon (top left), ultimate exclusivity by a Rolls Royce Phantom (top right), emotion and fun by a MINI Cooper (bottom left), or power and sportiness by a Porsche 911 Turbo (bottom right).

The fact that styling is as important for a vehicle's market success as its technical properties has been known since the 1930s. Until then, the sole driver for a vehicle's body style was optimization of technical parameters such as aerodynamic drag or stiffness – resulting in cars which consequently all looked more and more the same. Alfred Sloan, GM's chairman at this time, realized that the car had changed from a pure means of transportation to an article of fashion, and that it made sense to sacrifice a certain amount of its technical performance in order to obtain a distinctive and emotional styling. GM started offering an annual revisions of their models exterior style – and subsequently surpassed their big rival Ford in sales.

A more recent example for the importance of styling is Toyota's launch of its youth-oriented and U.S.-exclusive Scion brand in 2004. Two cars were brought to the market: The subcompact xA and the compact station wagon xB, both based on the Echo/Yaris platform. Planned sales volume for the xA was significantly higher than for the xB, but due to its distinctive styling and body concept the xB outsold the xA 2 to 1.

Fig. 7.5 Cars with strong exterior styling messages
(Sources: Lamborghini, BMW, BMW, Porsche)

While exterior styling is said to be responsible for "love at first sight", interior styling and value perceived should make this love everlasting. The fact that many people spend more time in their car than in their living room is one of the reasons for the current market trend towards premium-quality and distinctive vehicle interiors. As an example, Fig. 7.6 shows four completely different design approaches: Luxurious and comfortable in a Bentley Arnage Final Series 9, (top left), elegant and sporty in a BMW Concept CS concept car (top right), young and extroverted in a MINI Cooper S (bottom left), puristic and sporty in a Lotus Exige.

Aside from the exterior and interior, the engine and baggage compartments are also styled. The customer should experience the design language of the vehicle even when opening hood or trunk.

Fig. 7.6 Distinctive dashboard styles (Sources: Bentley, BMW, BMW, Group Lotus Plc.)

7.3.1.2 Ergonomics

A commonly accepted but very general definition of the widespread term ergonomics was adopted by the IEA Council in August 2000: "Ergonomics (or human factors) is the scientific discipline concerned with the understanding of interactions among humans and other elements of a system, ..." [10].

The central target of ergonomic vehicle design is the realization of both a spatial geometry and a man-machine-interface that allows driver and passengers to safely and comfortably operate all vehicle functions in all relevant situations – of course with the primary focus on the driver driving the car. Customer relevant properties regarding ergonomics are e.g.:

- Seating: Position, adjustability, variability, dimensions of the seats; static and dynamic seat comfort; foot well size; seat belt comfort.
- Dimensions: Leg room, knee room, hip room, elbow room, shoulder room, head room. Subjective feeling of spaciousness.
- Driving operations: Steering effort; pedal characteristics; handbrake, gear selector operation; accessibility of control elements; visibility of displays.

- Non-driving operations: Ease of entry and exit; passenger comfort; ease of loading and unloading; ease of refueling; ease of equipment installation (e.g. roof rack, trailer etc.).
- User interfaces for information, communication and entertainment systems: Intuitivity of operations, readability of displays, general management of the complexity caused by the increasing amount of information technology within cars.

Together with dynamic, acoustic and thermal comfort (see Sects. 7.4.1, 7.4.2 and 7.4.3), ergonomics contribute to the minimization of the driver's fatigue and discomfort and thus significantly influences his or her alertness and active safety. For this reason, many vehicle characteristics concerning ergonomics are subject to legal regulations (compare Sect. 7.1.1).

When optimizing vehicular ergonomics, one has to keep in mind that the occupants' physiognomy varies from market to market. Typically, Scandinavian men being among the tallest and far eastern women being among the shortest represent the extremes of anthropometrical height. Similar variations appear in factors such as width, waist line, hand size, reach, and weight. Achieving the same level of customer satisfaction worldwide is a big challenge for international car makers.

Another challenge to vehicle ergonomics arises from the fact, that people nowadays live longer and also drive their cars at an ever older age. These senior drivers have very high requirements concerning ease of operations whilst still expecting to use the full spectrum of its functions.

7.3.1.3 Practicality

Another customer relevant aspect of a vehicle determined by its design is its practicality, including both the vehicle's capability to store luggage and personal items and the dimensional practicality of the vehicle as a whole. Typical requirements from a customer's point of view are e.g.:

- General load space: Width, length, height; open / closed; total volume (e.g. number of VDA standard boxes)
- Load space for standard items: SAE standard luggage set; bicycles, golf bags, skis etc.
- Additional load space: Roof rack availability / capacity; towing hitch availability
- Size and position of trays, pockets, holders etc. for stowage of personal items
- Dimensional practicality: Overall length, width, height (especially in relation to e.g. maximum dimensions and weights of streets, garages, trailers, auto-trains etc.)

Practicality is one of the vehicle characteristics that are typically only perceived when not provided in certain situations: When the luggage for holiday travel does just not fit in the trunk, when there is not enough space to store a cell phone and sunglasses in the center console or when the car hardly fits in the customer's garage.

During the development process, the customer's requirements concerning practicality are represented by a *list of personal items* that indicates the items such as suitcases, sunglasses, sports equipment etc. for which storage space should be provided in a car. There are different lists with a growing number of personal items, representing different levels of practicality. The lowest level could e.g. be valid for a small two-seater roadster, the highest level to a multifunctional van. At the target agreement milestone, each new vehicle will be assigned a list of personal items that must be accomodated.

7.3.1.4 Sound Design

In addition to visual and haptic characteristics, a vehicle is also – and very directly – perceived by its acoustic properties, both in the cabin and the vehicle's environment. From a customer's point of view, vehicle acoustics have three different aspects:

- Vehicle sound as a differentiating design element (discussed in this section)
- Interior noise as an element of cabin comfort (discussed in Sect. 7.4)
- Exterior noise as a form of vehicle emissions (discussed in Sect. 7.10)

Particularly in the premium market segment, customers expect their car to have a distinct engine sound profile that fits the overall character of the brand and the vehicle type – naturally within the legal limits discussed in Sect. 7.10.4.1 [4]. Such consistent sound design comprises:

- Suppression of unwanted noises
- Acoustic feedback of driving dynamics (engine load, tire-road-friction)
- Permeability for external acoustic warning signals (horns, sirens etc.)
- Provision of an optimum acoustic environment for on-board entertainment and operation signals

Figure 7.7 depicts the preferred combinations of the general engine sound characteristic and the engine loudness during acceleration for different vehicle types. For cabin comfort reasons, the engine should not be audible at all in the cabin at constant speed. When under load, the engine should provide appropriate acoustic feedback to the driver.

In addition to engine sound, customer requirements also include appropriate sound design for all operations such as closing doors and flaps, lifting and lowering windows, actuating power seats, or operating switches. The sound should give clear feedback and reflect quality. Lowering a window e.g. should create a uniform, sovereign operating sound without any load related frequency fluctuation.

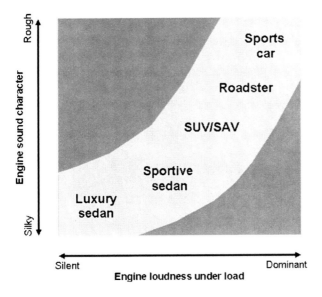

Fig. 7.7 Engine sound characteristics (Source: BMW)

7.3.2 Component and System Design

The ways and methods by which a new vehicle's design is worked out, evaluated and eventually determined, differ vastly among OEMs. As they reflect differing levels of emphasis placed on various product characteristics and functions, they are essentially a major factor of differentiation among the specific brands and OEMs. Even though this makes a generic model for automotive design processes difficult to create, there are elements and sequences that can be considered as general. For this book, the BMW Group automotive design process is taken as a specific example. A comprehensive description of this process is given in [9] and [11].

The design process is an integral part of the general PEP (see Chap. 1). BMW Group (BMW, MINI, Rolls Royce) designers use particular names for the phases of the PEP – according to the emphasis the design process has during the respective phase: During the project-independent product strategy, *advanced design* creates design visions for the future. As the main design focus during the early concept phase lies on understanding the design task agreement on proportions, this phase is called *understanding phase*. During the following *believing phase* (which is parallel to the late concept phase), the final exterior and interior design are selected. Eventually, the *seeing phase* incorporates detailing and realization of the design concept – as part of series development.

A special characteristic of the BMW Group automotive design process is internal competition. To obtain a very high level of design quality, up to eight different design teams work out different exterior and interior concepts – both during the understanding phase and the believing phase. In a gradual selection process, each concept proposal is evaluated. Eventually, a winning design is selected by the board of management. The upper level of the BMW Group automotive design process is illustrated in Fig. 7.8.

Fig. 7.8 BMW Group automotive design process (Source: BMW)

7.3.2.1 Advanced Design

As discussed in Sect. 3.1, corporate strategy and brand image represent the bookends between which project-independent advanced design takes place – the creative exploration of concepts for potential new vehicles. Stimulated by current customer trends, ideas, competition analysis, innovative materials and technologies, internationally distributed design studios create visionary exterior and interior design themes which eventually are incorporated in concept cars. As the rendering and the picture of the 2006 BMW Concept Coupé Mille Miglia[17] in Fig. 7.9 demonstrate, new or distinctive styling elements are particularly highlighted or even exaggerated on concept cars. But even if a concept car is never brought to series production, the public reaction it gets when displayed at international auto shows

---

[17] The BMW Concept Coupé Mille Miglia 2006 was revealed by BMW at the opening of the 2006 Mille Miglia oldtimer race in Brescia, Italy. Its design is inspired by the legendary BMW Mille Miglia Touring Coupé that won the Mille Miglia in 1940.

provides valuable feedback to management on whether the revealed design or function features will appeal to future customers or not.

Fig. 7.9 Concept car: Design sketch and drivable model (Source: BMW)

7.3.2.2 Understanding Phase

As a vehicle project starts, the innovations and design concepts identified during the advanced design stage are consolidated with the requirements of marketing (including the customer requirements stated above) and engineering in order to understand the design task and come to a consistent vehicle specification. During this phase, the specification is converted into sketches, *computer aided styling* (CAS) models and package plans that incorporate the general proportions of the vehicle as well as the position of the major components and occupants' seating positions.

Also, the major conceptual dimensional chains of the complete vehicle (e.g. right wheelhouse – right engine carrier – engine – left engine carrier – left wheelhouse) are determined by exterior and interior styling and are checked for feasibility during the initial phase (compare Sect. 3.1.1).

In parallel to the hand-made sketches, 2D and 3D CAS drafts are created. Usage of CAS systems allows the designer to develop rough surface geometries within a short period of time and to share this geometry data with all partners involved in the PEP.

To allow exterior and interior styling and packaging to work in parallel, *dimensional limit points* are stipulated. These points show e.g. the exterior designer how steep his roofline may descend without affecting the required rear passenger space.

The understanding phase can take up to one year. The results of the competing teams are typically represented by means of 2D sketches, 1:1 foam models or even 1:2.5 clay models. With the selection of one interior and one exterior proportion model at the end of this phase, proportions and character of the new vehicle are confirmed.

7.3.2.3 Believing Phase

Based on the selected proportions, the main focus of the believing phase is on the aesthetics of the vehicle, especially represented by means of surfaces and design lines. At the beginning of this phase, each design team expresses their idea of a design language with sketches and manual renderings like the one of a BMW 3 Series coupe depicted in Fig. 7.10).

Fig. 7.10 Manual exterior rendering (Source: BMW)

Next, contour lines are honed to perfection by applying adhesive tape of different width to 1:1 tape plans (see Fig. 7.11). This method allows a high flexibility to changes within the design concept, e.g. adding extra headroom or changing the dynamic appearance of contour lines.

As a next step, a clay model is derived from the tape drawings, marking the transition from 2D to 3D representation. Highly skilled specialists transfer the vehicle geometry meticulously from the 2D drawings to the clay, using high-precision 3D measuring devices and putty knifes. Covered with resilient foil, clay models give a first but very realistic impression of the new vehicles' – exterior and interior – visual and ergonomic appearance (see Fig. 7.12). In a way that would never be possible with CAS models, the 1:1 clay models widen the sensory perception and are the central platform for design work during the believing phase.

Interior design is followed correspondingly, only with a higher focus on materials and textures. Figure 7.13 shows different realization levels of an interior design concept.

Fig. 7.11 Taping of exterior contour lines (Source: BMW)

Fig. 7.12 Manufacturing of 1:1 exterior clay models (Source: BMW)

Fig. 7.13 Development stages of an interior design concept (Source: BMW)

In a process that is both intuitive and methodical at the same time, up to eight competing design teams are honing their 1:1 models through multiple try-test-discard-try loops, until they represent what the designer perceives as the perfect equilibrium between form and function. At the end of the believing phase, the final exterior and interior design models are carefully chosen by the board of management in a stepwise selection process [9].

7.3.2.4 Seeing Phase

The seeing phase is about "turning a sculpture that was refined by human hands into a product that can be reproduced by machines". Designers focus on getting the selected design concept realized as exactly as possible by series development and negotiate fractions of millimeters with their partners from engineering and production, e.g. whether the 2 mm bending radius of the trunk lid can be changed to 3 mm in order to improve manufacturability. This "fight for design" concerns all the changes that occur to vehicle development until its production is launched.

To make the design data available for all partners involved in the series development, the selected clay model is laser-scanned and in this way converted into a precise 3D CAD surface representation. This model is then used as the binding data reference for all following development [9].

7.3.2.5 Sound Design

Legal noise emission limits and cabin comfort requirements are the borderlines between which a vehicle's sound as an element of its design can be "styled". At the relevant low vehicle speeds, rolling noise and wind noise only play a minor role. The two sound sources that have potential for sound design are the engine and auxiliary actuators such as window lifters etc. (compare Sect. 7.4.2).

Sound design must start at an early stage in the development process. The relevant components that need to be optimized are:

- Engine/gearbox combination
- Exhaust system
- Electric actuators

The acoustic characteristics of these components are initially analyzed during the concept phase. After respective measures have been derived, these components will be pre-qualified when they are brought into complete vehicle integration in series development. Design measures taken for sound design must be in line with the often conflicting objectives concerning interior acoustics and vibrations as well as noise emission.

An approach to suit an engine's sound to the required vehicle characteristic is acoustic accentuation of engine orders. In an engine's sound spectrum indicating the sound level over frequency and engine rotating speed, the engine orders can be seen as characteristic lines through the origin. Figure 7.14 exemplifies how different sound characteristics can be achieved for a BMW six cylinder in-line through targeted modifications of the exhaust system: To create a cultivated, silky sound that suits a sedan, emphasis is placed on the main engine order (red line in the left picture); to create the sportive, sonorous, speed avaricious sound that suits a sports car, lower and higher level engine orders are accentuated (red lines in the right picture).

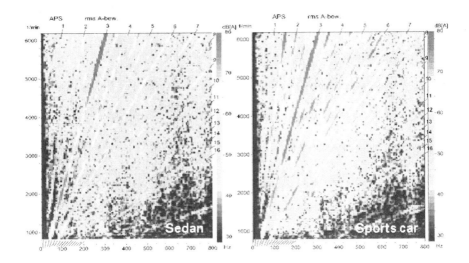

Fig. 7.14 Sound design through accentuation of engine orders (Source: BMW)

Selectable driving modes (such as *comfort* or *sport*) which are available in certain premium vehicles may – in addition to different suspension settings and gear-change-patterns – also offer different sound patterns. In this case, sound design elements such as resonators can be switched on and off according to the required sound characteristic.

7.3.3 System Integration and Validation

As stated above, design validation is an integral part of the design process, and different tools and methods are applied during the different stages.

7.3.3.1 Styling

For intermediate results and qualitative decisions, exterior and interior styling can be validated using rendered CAS models, visualized as pictures, 1:1 projections or virtual reality environments. Smoothness and harmony of surfaces are checked by means of computer aided tools (see Fig. 7.15).

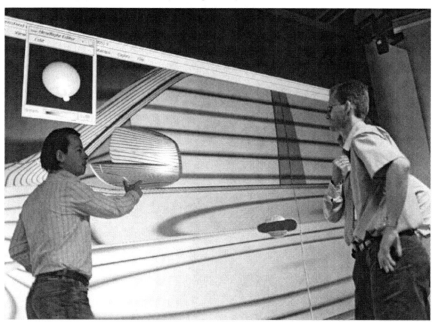

Fig. 7.15 Surface evaluation on virtual surface models (Source: BMW)

Photo-realistic representation of surfaces, materials and textures allow a very lifelike perception of exterior and interior design appearance without the necessity of expensive physical prototypes. Figure 7.16 shows a photorealistic exterior representation of the BMW Concept CS, Fig. 7.40 an interior detail.

Fig. 7.16 Photo-realistic exterior representation (Source: BMW)

When eventually decisions of consequence are made concerning exterior and interior design, it always relies on 1:1 models. Only realistic hardware allows the highly multi-sensual process of validating a car's proportions, surfaces, colors, odors, touch and feel etc. The number of 1:1 models used during this optimization and selection process directly influences the quality of the final design. Figure 7.17 shows the assessment of an early design model.

Fig. 7.17 Evaluation of an early design model (Source: BMW)

As styling strongly affects other complete vehicle characteristics, (e.g. manufacturing costs through manufacturability of body surfaces, acoustic comfort through aerodynamic drag, ride comfort through the space provided for suspension elements), it is developed in close cooperation between designers and engineers [12].

7.3.3.2 Ergonomics

Basic ergonomics criteria such as the adoption of healthy and efficient postures for the complete range of future users must be taken care of and validated in the design process as early as possible to prevent late and costly alterations [12]. In contrast to styling, conformity with ergonomics requirements can be thoroughly assessed at a very early design stage. Advanced simulation tools like RAMSIS or SAMMIE CAD allow precise investigations of driver and passenger ergonomics using virtual cars and scalable virtual manekins. Figure 3.5 shows a simulation to assess a ergonomics of the driver environment.

As vehicle ergonomics highly depends on individual physiognomy, validation must include a broad bandwidth of representatives of the future users. One method are *fitting trials*, during which seating positions, the position of displays and control elements etc. are evaluated by 20 to 30 individuals in an experimental driving rig [12]. A common method to evaluate ergonomics for the largest possible variety of occupants and make these data available for future vehicle projects are questionnaires that are sent out to vehicle owners several weeks after purchase.

7.3.3.3 Practicality

To ensure that a vehicle concept provides the required space to store the items specified in the list of personal items at locations convenient for driver and passengers, virtual representations of these items are positioned in the virtual car and included in the virtual car process (compare Sect. 4.2). The complete set of virtual personal items is shown in Fig. 7.18.

While virtual validation can ensure the basic possibility of storing the required items, realistic fitment of e.g. non-rigid items such as golf bags can only be checked using hardware representations of the respective storage area. As these requirements can determine concept-critical dimensions (such as e.g. the maximum width of the rear trunk opening), their fulfillment must be confirmed as early as possible in the project. For this purpose, concept mock-ups as shown in Fig. 7.19 are used, embodying the storage area as well as critical borders (in this case the rubber sealing of the rear trunk lid). Based on concept CAD data, these concept mock-ups can be quickly built at low cost and allow realistic evaluation of fitment, ergonomics and space.

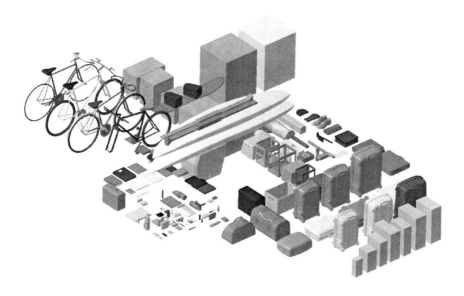

Fig. 7.18 Virtualized personal items (Source: BMW)

Fig. 7.19 Early concept mock-up for trunk practicality evaluation (Source: BMW)

Figure 7.20 shows the validation of standardized trunk capacity in simulation (left) and reality (right) according to VDA directive 210.

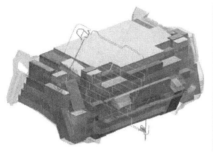 

Fig. 7.20 Measurement of formal trunk capacity (Source: BMW)

7.3.3.4 Sound Design

One of the challenges in evaluating the individual perception of the interior sound of a given vehicle is, that the human acoustic memory only lasts for about 20 sec – which makes direct comparison of competing models or conceptual alternatives close to impossible. To make the acoustic perception of driver and passenger reproducible and thus enable time-independent evaluation, the binaural sound experience in question is recorded via an artificial head – an acoustic measuring system consisting of auditory canals and microphones disposed in a human torso (see Fig. 7.21). The recorded sound sample then can be replayed in a studio situation equalized by correction filters and be precisely evaluated by one or more sound engineers [13]. A comprehensive sound library allows direct comparison with former models or competition vehicles.

In order to make the subjective human perception of sound measurable and comparable, three psychoacoustic qualities of sound perception are identified which then can be measured by physical variables [14]:

- Loudness: Describes the perception of a sound by an occupant. Depends on the pressure level, the frequency and the duration of the sound. Measured variables are e.g. sound pressure level [dB] or speech comprehensibility / logatom comprehensibility [%].
- Dynamics: Describes the dynamic behavior of the sound, such as development of the engine sound when increasing the engine speed or load. Measured variables are e.g. frequency change over time [Hz/s] or sound pressure level increase over time [dB/s].
- Timbre: Perceived quality and color of a sound, depending primarily upon the sound spectrum (overtones) and waveform. The measured variable is e.g. the maximum sound pressure levels for each engine order [dB].

Fig. 7.21 Vehicle sound recording using an artificial head (Source: BMW)

7.4 Cabin Comfort

Cabin comfort denotes the overall feeling of well-being of a vehicle's driver and passengers which is perceived through all of their senses. A highly individual and subjective experience, it comprises *vertical dynamics, thermal comfort, acoustic comfort* and *value perceived*. As cabin comfort is an enabling or limiting factor to the driver's alertness, cabin comfort is one of the elements that determine a vehicle's level of active safety.

7.4.1 Riding Comfort

7.4.1.1 Legal and Customer Requirements

Riding comfort for driver and passengers is determined through frequency and amplitudes of – mostly vertical – oscillations which are induced in their bodies during the drive. The human body reacts very sensitively to certain frequencies, and exposure to these frequencies can lead to all levels of discomfort including immediate motion sickness [12]. Typical customer expectations concerning riding comfort are:

- Total absence of annoying or sickening vibration phenomena for all vehicle occupants
- A body movement that is appropriate for the vehicle's design
- Direct perception of the road by the driver
- Enabling handling and agility
- Safe and comfortable seating under all driving conditions

In addition to vertical body oscillations experienced through seat and floor, occupants are exposed to lateral and longitudinal vibrations and part specific vibration phenomena – such as torsional vibrations of the steering wheel or vibration of the gear shift lever. Figure 7.22 shows different customer relevant vibration phenomena and their respective areas of acceleration and frequency.

Requirements for ride comfort vary greatly from market to market. European customers e.g. usually drive shorter distances at higher speed on excellent roads and hence prefer tight suspension and upholstery in order to get a more direct contact to the road. In the U.S., cars are usually driven at lower speed over longer distances; drivers and passengers hence place higher importance on soft suspension, maximum damping resulting in a high level of comfort.

As vertical dynamics determine the vehicle's wheel-street contact at any moment, they are also a physical foundation – and a limiting factor – for longitudinal and lateral dynamics (compare Sects. 7.6.1.1 and 7.6.1.2). And with ride comfort only indirectly influencing vehicle safety or emissions, there are no legal requirements concerning ride comfort.

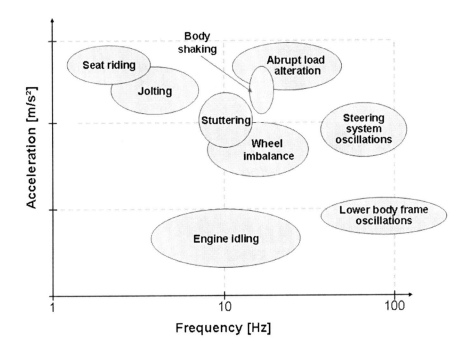

Fig. 7.22 Customer relevant vibration phenomena (Source: BMW)

7.4.1.2 Component and System Design

There are two physical sources for vertical oscillations (along the z-axis and around the x- and y- axis) of the body:

- The relative movements of the wheels along the z-axis according to the dynamic contact between road and wheels, induced in the body via the chassis and spring-damper-system
- The unbalance of rotating and oscillating elements of the engine and the drivetrain, induced in the body via engine and drivetrain mounts.

The overall behavior of the complete vehicle oscillation system, is determined by the visco-elastic properties (stiffness and resonance frequencies) of the following elements:

- Sprung masses (body and trim, engine, drivetrain etc.)
- Unsprung masses (wheels, tires, wheel carriers, brakes etc.)
- Rotating and oscillating masses (shafts, pistons etc.)

- Passive damping elements (upholstery, spring-damper-systems, rubber mounts, mass dampers, absorbers etc.)
- Active control systems (roll stabilization, damper control systems, controllable engine mounts etc.)

Figure 7.23 shows the visco-elastic system that determines vibration behavior of the complete vehicle.

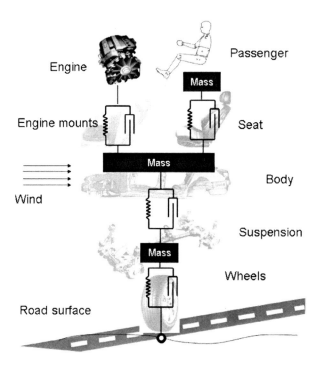

Fig. 7.23 Visco-elastic vehicle system

The key conceptual design approach to optimize the complete vehicle in terms of vibration comfort is to make sure that the resonant frequencies of especially body and chassis parts are far from the excitation frequencies so that none of these elements goes into resonance. A rule of thumb here is to keep frequencies apart by a factor of $\sqrt{2}$. Figure 7.24 illustrates this approach: The wheels are excited by rolling over a rough road. As these vibrations are transmitted to the body, wheels and chassis components vibrate according to their characteristic frequency. To avoid resonance and uncontrolled oscillation, the resonant frequencies of the body must be distant enough from the peaks of the exciting frequency. Thus, the peak level of the oscillations perceived by the occupants in the cabin is acceptable. If however the resonant frequencies of the body became lower – e.g. by enlarging the roof cutout or the rear end opening – the excitation frequency transmitted by the wheels might now match the lower body resonance frequency leading to an annoying low frequency humming coming from the rear end of the vehicle.

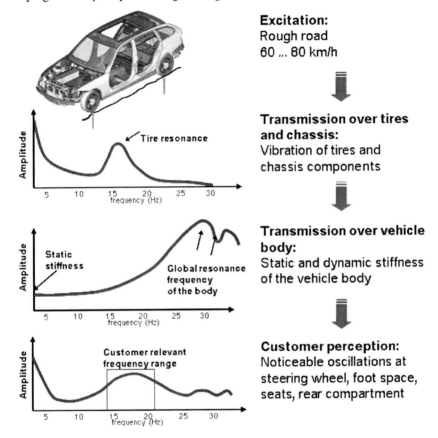

Fig. 7.24 Transmission of vibrations through tires, chassis and body (Source: BMW)

If the required distance between exciting and resonant frequency can not be achieved (e.g. when an initially unplanned new engine is integrated into an existing vehicle concept), possible correctional measures are to increase the stiffness of e.g. the body, or shift of the resonance frequency by means of struts or the application of absorbers.

Optimization of ride comfort typically conflicts with vehicle agility targets. Design approaches to solve this conflict include [4]:

- Load leveling, e.g. through inflatable and deflatable air springs
- Adjustable damper systems
- Active springs or stabilizers

7.4.1.3 System Integration and Validation

As even minor changes in material or geometry can increase or reduce a component's resonant frequency, ride comfort as a complete vehicle characteristic is very sensitive to variation of its parameters. For each configuration made up by variants of body, engine/gearbox combination, spring-damper system, seats and even tires, the behavior of the complete vehicle might be completely different, and thus requiring specific integration. Tuning the car in order to meet the riding comfort requirements is usually done by specifying a distinctive spring-damper-system, adding and adjusting passive damping elements and adjusting the control parameters of active control systems.

To evaluate and subsequently optimize the vehicle concept in terms of riding comfort at an early stage, the dynamic behavior of the complete vehicle can be simulated using *multi-body system* (MBS) models like the one shown in Fig. 7.25.

The next step in validating and optimizing riding comfort is the assessment of the vibration behavior of the complete vehicle on a hydro-pulse test rig that realistically simulates the excitations from driving the car without having to leave the development site. The vibrations driver and passengers are exposed to during these tests are measured using vibration dummies which are able to monitor vertical and horizontal oscillations by frequency and amplitude over time. By adding and removing weight, the vibration dummy can represent different kinds of occupants. Figure 7.26 shows a MEMOSIK V vibration dummy [15].

The final validation of riding comfort however requires on-road vehicle tests. Highly skilled test drivers who have a clear picture of the relationships between the perceived phenomena and the technical parameters that affect them are required in order to achieve the high level of riding comfort that is required for premium vehicles. Standardized test track surfaces like potholes, sinus waves or cobble stone (see Fig. 7.27) allow the reproduction and systematic elimination of unwanted vibration phenomena.

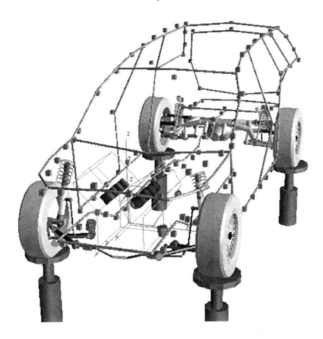

Fig. 7.25 Multi-body simulation of vehicle oscillations
(Source: BMW)

Fig. 7.26 Vibration measurement dummy (Source: WÖLFEL)

Fig. 7.27 Standardized ride comfort test tracks (Source: BMW)

7.4.2 Acoustic Comfort

Even though the underlying physical effects are similar, audible oscillations (noise) and non-audible oscillations (vibrations) are perceived differently by a vehicle's occupants and thus need to be treated differently over the development process.

7.4.2.1 Legal and Customer Requirements

While the physical sources for exterior and interior noise are the same, customer requirements and ways to handle it differ significantly. The major requirement concerning exterior noise is to remain below the legal noise emission level (compare Sect. 7.10.4.1) and at the same time to offer a sound design that suits the character of the vehicle and the brand (compare Sect. 7.3.1.4). In contrast to this, there are no legal requirements for the noise level in the cabin. However, interior noise is highly customer relevant and perceived by driver and passengers as a major aspect of travel comfort and active safety. Hence, especially comfort oriented premium cars have tight self-imposed interior noise targets.

7.4.2.2 Component and System Design

Acoustic comfort requires early conceptual design to avoid later addition of costly (and heavy) means of sound proofing (i.e. isolators and dampers) in order to meet the requirements. Optimization of a vehicle system regarding its interior acoustics requires identification and optimization of all noise sources – structure borne and airborne – and their transfer pathways into the cabin and further to the occupants' ears. Essentially, there are three sources of noise emission in a cruising vehicle:

- Rolling noise (rolling of the tire on the road)
- Drivetrain noise (engine, intake, exhaust, transmission, driving axle)
- Wind noise (air flow around the vehicle and through its cooling air system.

Table 7.2 lists the components that represent the main noise transmission pathways into the cabin and the locations for potential soundproofing measures. Figure 7.28 shows how the interior noise level of a vehicle can be improved by acoustic optimization of the body structure as the main transfer path for all types of noises (compare Sect. 7.4.1.2).

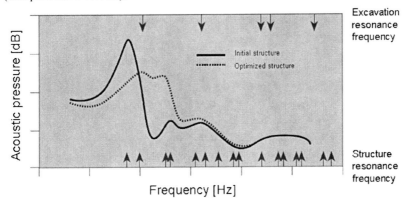

Fig. 7.28 Improvement of the interior noise level through structural measures (Source: BMW)

Over the last 20 years, transmission of rolling and drivetrain noise into the cabin has been dramatically reduced, partly by acoustic improvement of the components and partly by improvement of the transmission pathways through absorption or isolation. As a consequence, wind noise has become the dominant noise source at driving speeds above 100 km/h – both in interior and exterior noise. To qualify single exterior components, their individual contribution to the vehicle's wind noise is analyzed in an aero-acoustic wind tunnel, where wind noise can be decoupled from drivetrain and rolling noise. Figure 7.29 illustrates the general design of an aero-acoustic wind tunnel in a side section; Fig. 7.30 shows the measurement chamber with the vehicle under test in front of the air outlet.

Table 7.2 Sources and determining parameters for interior noise

Source	Parameters determining the interior sound level	Locations for soundproofing elements
Wind noise	Body style Exterior mirror shape Glass panel thickness Door concept Packaging concept Sealing concept	Side panel, headlining, rear / trunk, underbody
Rolling noise	Tires Chassis / suspension concept FWD/RWD Bearing isolation Body stiffness	Wheelhouse, underbody, rear / - ventilation
Drivetrain noise	Engine air-borne noise Engine structure-borne noise Suction noise Muffler noise DU engine/drivetrain Engine/transmission mount Transmission Propshaft/center bearing Rear axle /differential HVAC compressor / belt drive	Hood, underbody, rear / - ventilation
Noise from auxiliary equipment	ABS/DSC/CBC Power steering Auxiliary heating Heating cycle valves HVAC flaps Windshield wipers Fan Fuel pump Window lifters Sunroof actuator Central locking Mirror adjustment Seat adjustment	

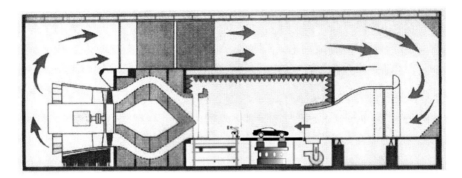

Fig. 7.29 Operating principle of an aero-acoustic wind tunnel (Source: BMW)

Fig. 7.30 Aero-acoustic wind tunnel (Source: BMW)

For the actual component analysis, all exterior edges that would create wind noise (e.g. body gaps, window seals, door handles etc.) are neutralized with adhesive tape first. Then, the fully taped vehicle is positioned in the wind tunnel and a pure wind noise spectrum is recoded. Now, the tape is removed from the first component. A new noise spectrum is recorded, and the deviation from the first one represents the aero-acoustic contribution of that component. Hereafter, the component is taped again, and the same procedure is repeated for all other relevant components. At the end of this procedure, a spectral analysis like the one in Fig. 7.31 is created from the obtained data, in which each color represents one component. This analysis shows clearly that wind noise not only appears as high frequency hissing but even more as low frequency booming of almost all body and trim parts – a fact that is usually underestimated.

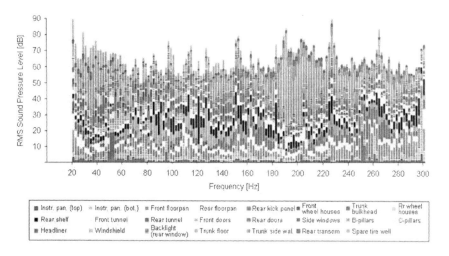

Fig. 7.31 Aero-acoustic vehicle sound spectrum (Source: BMW)

FE-based modal analysis (see the model shown in Fig. 7.32) allows identification of structural resonance frequencies and helps conceptual optimization of structural dynamics early during the PEP and – later on – optimization of noise reduction element layout.

While passive noise reduction uses absorbing or isolating elements, active noise reduction technology uses an acoustic effect called cancellation by interference: A sound wave is neutralized by adding a second sound wave with the same amplitude and the opposite polarity to it. While this effect is impressive in headphones, there has been no convincing application of active noise reduction in cars so far.

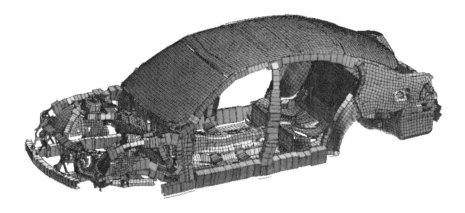

Fig. 7.32 FE-based modal analysis of the vehicle body (Source: BMW)

7.4.2.3 System Integration and Validation

Simulation systems such as *LMS Virtual.Lab Interior Acoustics* allow system-level interior acoustics validation. With given geometry, material data, boundary conditions and excitations, the acoustic structural responses can be calculated and visualized and thus be considered during conceptual design without any hardware testing [16].

Validation methods for the assessment of interior noise are to the same as those for sound design which have already been discussed in Sect. 7.3.3.4.

7.4.3 Thermal Comfort

7.4.3.1 Legal and Customer Requirements

Customers require a cabin climate that is comfortable and safe both for the driver and passengers, independent from the actual weather conditions. Functionalities for thermal cabin comfort however vary greatly: While low priced vehicles may be equipped only with a basic heating and defrosting unit, air conditioning has almost become a standard in middle class vehicles and above. An increased level of comfort can be obtained by additional features such as air purifying units, seat heating or cooling or individual temperature and airflow settings for each occupant or by effective pre-conditioning through parking ventilation, heating or air conditioning.

The main performance criteria by which a *heating, ventilation and air condition system* (HVAC) is perceived by the customer are:

- Agreeability: Pleasant temperature, air humidity, temperature layers (cold head, warm feet), draft-free and noiseless ventilation, air quality (particles, smell, oxygen)
- Dynamics: Fast heating up, cooling down, defrosting, demisting
- Stability: Bandwidth within which the selected temperature is controlled

7.4.3.2 Component and System Design

Perception of thermal comfort in the vehicle cabin is highly subjective, and the parameters that influence and steer this perception are plentiful (see Fig. 7.33).

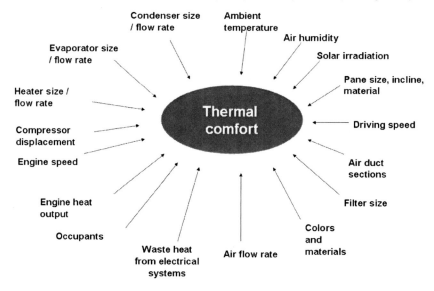

Fig. 7.33 Factors influencing thermal cabin comfort

From the four subsystems that make up an HVAC system (heating unit, cooling unit, ventilation system and control system), it is especially the ventilation (consisting of fan, heat exchangers, control flaps and air ducts) which is a highly integrated system, with geometric, thermal and acoustic constraints. Taking the HVAC system of the current BMW 7 Series as an example, Fig. 7.34 depicts the air flow scheme representing the system's functionality, and Fig. 7.35 the corresponding 3D-layout of air ducts which represent the systems geometry in the cabin.

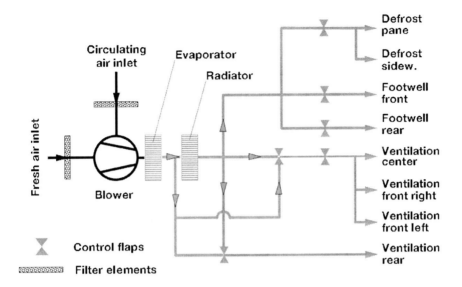

Fig. 7.34 HVAC system air flow scheme (Source: BMW)

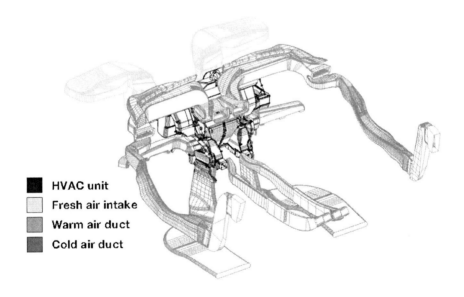

Fig. 7.35 HVAC system air duct layout (Source: BMW)

The quality of air ducts can be measured by:

- Pressure loss along the duct
- Air-flow distribution at air vents
- Air-velocity distribution on window panes (defrost)
- Response to temperature changes
- Noise

These parameters are all determined by the geometry of the ducts: The length, the section shape and the gradient of the shape along the duct. Generally, ducts should be circular or at least rectangular, without sharp edges, abrupt section changes, corners, bellows etc.

7.4.3.3 System Integration and Validation

Simulation tools allow early conceptual validation of customer relevant thermal comfort characteristics. Heating / cooling performance, air flow, velocity distribution, temperature distribution, heat-up (passive) and cool-down performance or window defrost and ventilation can be predicted with about 80% exactness. Figure 7.36 shows an x-z cross-sectional simulation of the cabin airflow, indicating temperature, direction and speed of the air-flow. Required input data are the geometry of the cabin interior and the air ducts, the cockpit package and the specification data of the HVAC system components (air mass flow, heating/cooling performance etc.).

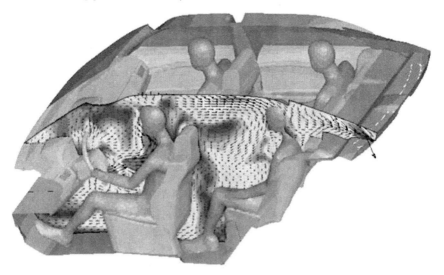

Fig. 7.36 2D cabin air flow simulation (Source: BMW)

The challenge in quantifying the real thermal behavior of the cabin lies in driving the car in reproducible environmental conditions. This can be achieved by driving on rolls in a climate test booth. Figure 7.37 illustrates the operating scheme, Fig. 7.38 shows an actual test booth.

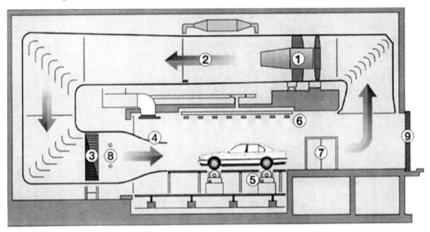

① Blower   ② Airstreams   ③ Heat exchanger

④ Vent   ⑤ All-wheel-drive roll   ⑥ Solarium

⑦ Refueling system   ⑧ Air-humidity controller   ⑨ Test booth gate

Fig. 7.37 Climate test booth scheme (Source: BMW)

Fig. 7.38 Climate test booth (Source: BMW)

To realistically validate thermal comfort in the cabin, vehicles need to be tested under real extreme weather conditions. To bypass seasonal restrictions, car manufacturers use proving grounds on both the northern and southern hemisphere. Hot climate testing e.g. can be carried out in the summer in Greece and in the winter in South Africa. Figure 7.39 shows a winter test in Sweden (left), and a hot climate test in the South African desert (right).

Fig. 7.39 Extreme climate testing (Source: BMW)

7.4.4 Value Perceived

7.4.4.1 Legal and Customer Requirements

Aspects of comfort determined by interior design have already been discussed in Sect. 7.3.1: Shapes and contours of interior surfaces, placement of controls and displays, driver and passenger ergonomics or practicality through trays and pockets are distinctive design characteristics of every vehicle. But to provide an overall feeling of value, a vehicle's interior must offer more than this; it must offer an integral experience of exclusiveness, authenticity and comfort. In the automotive premium segment, customers expect to perceive this level of value through all their senses. The *value perceived* is the interplay of visual, haptic, olfactory and acoustic properties of surfaces, gaps, displays and control elements. Typical customer requirements concerning the value perceived are:

- Surfaces: Flowing, harmonic running of highlights on different materials achieved by perfect surface body contours. Homogeneous and consistent surfaces and colors. Consistent material concepts. Authentic look and feel of distinctive materials. Scratch insensitivity.
- Displays and control elements: Consistent concept for control elements and displays – including nighttime appearance. High readability and graphic coherence of displays. Clear and consistent functional haptics and acoustics of control elements.

- Joints and gaps: All joints are equal in width and are placed in parallel regardless of the view angle. Directly related components fit perfectly.
- General impression: No other components should be visible through grills, depressions, joints and cavities.
- Operation sound: Pleasant and distinctive acoustic feedback from control elements and actuators (e.g. window lifter).
- Odor as an integral experience in the cabin: Olfactory perceptibility of natural materials such as leather or wood. Total absence annoying smells (such as polymer evaporation).

In the future, high standards in terms of value perceived won't be restricted to premium vehicles. There is a clear trend, that customers expect a consistent perception of value and quality also in the mass market [4].

7.4.4.2 Component and System Design

Design for value perceived means caring about details in a consistent way. Colors, surfaces, odors and sounds should all together follow a common pattern that represents the character of the brand and the specific vehicle. To achieve this, OEMs usually define design rules which must be followed by every designer responsible for interior parts or systems. Examples for such design rules are:

- Surfaces that look the same must also feel the same.
- Not more than 3 different surfaces concurring anywhere in the car.
- Comparable operating elements give comparable feedback.
- Directly related component geometries have to create a balanced and exactly aligned appearance.
- Fixing elements are generally invisible.
- Characters in displays are coherent in terms of font, type size, font style, color, positioning and implementation of day and night design.
- Operating noises such as shutting a door or adjusting the seat must match defined sound patterns.

7.4.4.3 System Integration and Validation

Today, powerful rendering algorithms allow a very realistic representation of 3D CAS models which allow designers to quickly pre-evaluate different design variants at relatively low cost. Shadows, light effects, textures etc. can be created in a quality that is equal to photography. As an example, Fig. 7.40 shows a photorealistic picture of the gear-shift lever of a BMW 6 Series that allows realistic evaluation of the visual appearance of different surfaces including light effects and shading.

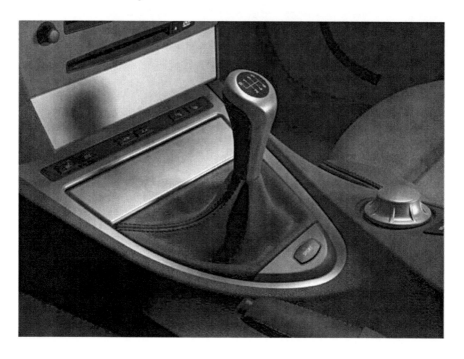

Fig. 7.40 Photo-realistic representation of interior parts (Source: BMW)

In contrast to the visual appearance, operating sounds, haptic perceptions or odors can only be evaluated in a real vehicle with surfaces, materials, operating elements and displays that represent series production stage. Here, the challenge is that the respective design features such as grain are only available very late in the development process.

7.5 *Infotainment*

In automotive engineering, the term *infotainment* summarizes E/E systems that are intended to meet information, communication and entertainment requirements of the vehicle's driver and passengers. While complete vehicle characteristics such as agility or passive safety are supported by electronics, infotainment is the area where electronics have become directly customer relevant features themselves. Infotainment systems include navigation systems, warning systems, and a variety of telecommunications devices and services that deliver information and entertainment to drivers (e.g., email, internet access, and location based information such as gas stations, restaurants, traffic and weather). Automated driver assistance systems include collision warning, adaptive cruise control, lane departure warning, lane change aids, and parking aids. As these functions grow ever more intertwined, the distinction between infotainment and driver assistance systems is becoming increasingly blurred [17].

7.5.1 Legal and Customer Requirements

Over the past 20 years, no area of vehicle functionality has developed as rapidly as infotainment: Only two decades ago, middle class cars were usually equipped with a radio and cassette tape player. Traffic information was available from local radio stations. Later, the first luxury cars were equipped with a built-in mobile phone – and could be easily spotted by their impressive rod antennae. Following were the first navigation systems – initially also only in the luxury segment – and rarely calculated a route that was the shortest or fastest.

With global traffic conditions moving to ever lower average speeds and awareness for emissions and energy consumption caused by motorized traffic ever increasing (compare Sect. 7.10.1), infotainment functions are becoming a more and more important aspect in the customers' purchase decision. Today, CD/DVD players are standard, MP3 players are available even for smaller vehicles. Integrated navigation systems with dynamic route generation and enriched 3D visualization or video playback for rear seat entertainment are features that distinguish one premium vehicle from another one. Sophisticated sensors such as cameras or radar devices allow additional information functions like lane departure warning or dynamic distance control. And above all, customers expect a *human machine interface* (HMI) that allows the driver and the passengers to intuitively access and control all of the available features.

The most important question for the customer obviously is which of all the functionalities his or her car offers. Table 7.3 shows typical expansion stages of infotainment systems.

Table 7.3 Functionalities of infotainment systems

	Current	**Future**
Information	• Traffic information	• Online traffic information
	• Status of vehicle components / liquids	• Preplanned routes
		• Car-to-car information
	• Navigation verbal / arrow / display	• Text to speech
	• Driver warning systems (lane departure, distance …)	
	• E-Call	
	• Internet access	
Communication	• Cell phone	• Voice over IP
	• PDA	• On-board email
		• Access to personal data (address book, calendar, notepad etc.)
Entertainment	• Radio	• Netradio
	• Digital broadcast (e.g. DAB, DVB-T, SDARS)	• Music download
	• CD/DVD player (conv. audio,	• Video on demand
	• Interface for external audio players (MP-3, solid state memory …)	
	• Video playback	

Apart from the range of functionalities, performance or speed is essential; e.g. for navigation systems, the time needed for route calculation is critical when the driver is ready to go but has to wait for the navigation system to tell him in which direction. When a car is driven in foreign countries, radio, navigation, phone, traffic information etc. should work – of course in the selected operation language.

Being the physical means by which the driver and passengers operate and communicate with infotainment and other electrical systems, the design of the HMI is highly customer relevant – and often a decisive factor for purchase. In times when technical differences between cars competing in the same segment are steadily shrinking, the HMI is one of the few areas, where different manufacturers today follow completely different approaches to satisfy the customers' needs. The main characteristics are the way functions are called or data input is given and how feedback or information is received from the system.

Another customer requirement is the previously mentioned integration of personal *consumer electronics* (CE) devices such as cell phones or *personal digital assistants* (PDAs). Seamless integration means, that the customer can use his or her personal devices – which are usually bought independently from the vehicle – via the HMI and has immediate access to stored data like phone numbers or addresses to use for communication or navigation.

Legal requirements concerning infotainment systems mainly focus on the minimization of driver distraction caused by the usage these systems. As e.g. studies have shown that the impairments associated with using a cell phone while driving can be as profound as those associated with driving while drunk [18], most countries ban the usage of hand held cell phones while driving and thus indirectly require provision of technical means that allow the integration of cell phones in the vehicle's HMI. Another important aspect of driver distraction is the position of displays and controllers. As a rule of thumb, controllers that cannot be operated blindly and displays should not be positioned below a viewing angle of 30°.

In the future, the possibility to exchange data with the vehicle's environment over wireless broadband internet connections will increasingly allow the utilization of off-board computing power (e.g. for navigation route planning) and access to off-board data (e.g. music files stored on the home PC) and thus open up a new dimension of functionalities for infotainment systems.

7.5.2 Component and System Design

The ever increasing demand for more and better infotainment functionalities has changed the overall architecture of vehicle infotainment systems dramatically over the recent past. Figure 7.41 indicates the different levels of integration these systems have gone through in automotive development. It also shows the increase in communication with off-board data and program servers.

While separate functions – such as radio, navigation or climate control – used to be realized in the vehicle by separate devices, today they are usually integrated into one central unit (so-called "head unit"), consisting out of five main elements:

- A main control unit (motherboard with CPU, solid-state memory units, graphic and sound processors)
- Software for the user menu and all integrated infotainment functions
- Wireless interfaces for data exchange with central information servers
- Input control elements
- Displays

	Integration level	Off-board data exchange	
1970	Replacement of mechanic functions by simple and separate E/E components	None	
1980	Enhanced functionality through integration of separate E/E components	None	
1990	New, additional functions not feasible without use of E/E through enhanced integration of E/E components	Little (9.6 kBit, SMS, WAP)	
2000	Integration of all E/E functionalities in a central unit ("E/E system")	Medium (115 kBit, GPRS, internet)	
2010	Integration of on- and off-board functionalities. The vehicle becomes part of a superior "system of systems"	High (2MBit, UMTS, internet – Java)	

Fig. 7.41 Integration levels of infotainment systems (Source: BMW)

The main design parameters for head units are:

- Access to the different functions: Through the user menu using a central operating device – or through distinct push-buttons
- Location of the central operating device
- Intuitivity and consistency of the user menu
- Size, quality and position of the central display
- Availability of additional displays such as monitors for rear set entertainment or a *head-up display* (HUD) to project relevant information directly in the driver's viewing angle and minimize distraction
- Variability of the operation language
- Playback quality of speech, music and video information

As mentioned before, manufacturers' different interpretation of customer requirements and different technical solutions lead to significantly different head unit concepts. Figure 7.42 shows four different design approaches: One single central controler for all functions in a BMW M6 Convertible (top left), fast hard keys arranged around the central display in a Porsche Cayenne GTS (top right), a touchscreen as a combined in- and output device in a Jaguar XF (bottom left), and a 3D display sphere in the MINI Crossover Concept concept car (bottom right).

Fig. 7.42 Different head unit layouts (Sources: BMW, Porsche, Jaguar, MINI)

Another design decision to be taken by the OEM is which infotainment functions should be integrated in the vehicle – and thus made proprietary, and for which an open interface should be provided that allows the customer to connect third party products to the vehicle system. Table 7.4 shows three different levels of integration:

Table 7.4 Integration levels of infotainment functionalities

Open interface	MP3 player Cell phone PDA	Personal devices with personal data – usually purchased independently from the vehicle – expected to be operable through the vehicle's HMI.
Integration	Radio CD / DVD player Navigation	Customers expect fully integrated solution in the premium market. Integration makes the system proprietary and prevents theft. Only in the lower-price segment, after sales products can be added to the vehicle after production.
Centralization	Traffic information (TMC) Address book Music / video on demand	Data is provided and processed through an external data and program server with which the vehicle communicates via a wireless broadband connection.

While vehicles are usually redesigned every 7 years, e.g. a cell phone is replaced by its successor on average after 6 months. Thus, the big challenge in integrating CE devices into a vehicle is to design the car and its interface in such a way that even a cell phone that will be developed 5 or 6 years after the car has been brought to the market can be used with the car. A customer who just purchased a brand new vehicle for some 50,000 dollars would probably never accept a requirement to use a cell phone with it that is 10 generations old.

To define and agree upon standard interfaces between the mobile devices and automotive E/E systems is the goal of CE4A,[18] a working group within the VDA. CE4A is organized into eight expert groups, which are engaged in different areas of expertise: *Phone, Media, PIM, Navigation, Terminal Mode, Standard Connector, Legal Form,* and *Reference Platform* [19]. An activity of the Phone Expert Group e.g. is the adaptation of the Bluetooth standard to the requirements of the automotive industry.

7.5.3 System Integration and Validation

While the general aspects of E/E system integration have already been discussed in Sect. 5.2.6, a specific challenge in integrating infotainment systems into a complete vehicle is – aside from the aforementioned integration of CE devices – the mutual dependency of electronic (functional) and non-electronic (geometrical/mechanical) properties: The central operating unit of the HMI e.g. must fulfill requirements in terms of functionality (give the intended input signals), ergonomics (position related to the driver, required movement, acoustic or visual feedback

---

[18] Current members of CE4A are Audi, BMW, Daimler, Porsche, VW.

from operation), value perceived (haptics, surface appeal), and styling (shape, color). As a consequence, requirements from all of these complete vehicle characteristics have to be specified and tested when functionally validating the vehicle's infotainment system.

Another specific aspect of integration and validation of infotainment systems lies in the dependency of the functionalities on local infrastructure and settings. Obviously e.g. neither the road network, nor the signals emitted from GPS satellites, nor the mobile phone network of Europe can be simulated in a development centre in the U.S. or Japan. Thus – in contrast to e.g. CD players or PDAs – the functionality of navigation systems or cell phones for the European market can only be tested in Europe.

7.6 Agility

7.6.1 Legal and Customer Requirements

Agility is the vehicles ability to transform the driver's commands – given through steering wheel, gas pedal and brake – into the intended movement. It describes the way, the speed and the extent to which a car can be accelerated, decelerated and steered by the driver under different driving conditions, and how the resulting movements are perceived by the driver and the passengers. Even though agility is commonly interpreted as a vehicle characteristic only needed by sporty drivers, it means much more than the ability to drive a car through curvy mountain roads at high speed. Agility also determines how safely a car can be handled in critical driving situations and thus is an essential element of active safety.

From the customers' point of view, a vehicle's dynamic behavior has three different aspects – differing in the direction in which forces are applied to the vehicle, its passengers and luggage:

- Longitudinal dynamics: Acceleration and deceleration performance, response of the car to acceleration and deceleration
- Lateral dynamics: Response of the car to forces crosswise to the direction of travel, induced by cornering
- Vertical dynamics: Response of the car to vertical forces, induced by the tires rolling off the road geometry and relative movements of the engine

Handling and comfort are the two – conflicting – characteristics of a vehicle's dynamic behavior. The vehicle concept determines where on the scale between a comfortable sedan and a race car the dynamic properties should be. A roadster for example certainly requires a different ratio of sportiness and comfort than a sedan or a mini van. And even within a given vehicle concept, derivates must be differentiated by their dynamic behavior. For example, the customer of a BMW M3 has definitely different expectations regarding the vehicle's dynamics than the one of a BMW 320i – although both cars are derivates of the same model.

7.6.1.1 Longitudinal Dynamics

Longitudinal dynamics of a vehicle are experienced by the driver during straight-ahead driving. Especially during acceleration, deceleration and load alteration, but also when driving at constant speed or up- and downhill. Here, typical customer requirements relate e.g. to:

- Performance (value and availability of engine power and torque)
- Traction (for acceleration and deceleration)

- Stability (for vehicle handling)
- Braking distance (on various surfaces)
- Body movement during load alteration

With regards to vehicle performance, the most important customer relevant characteristics are *acceleration* (given in seconds needed from 0 to 100 km/h or 0 to 60 mph), *nominal power* (given in hp or kW), *top speed* (given in km/h or mph), and *ampelstart[19] performance* (given in m or yards covered in 4 sec). Other requirements which are not as directly measurable are *sovereignty,* meaning effortless acceleration as e.g. required when passing another car, and *sportiness*, e.g. through a short response time during tip-in or an engine capable of high rpm's.

   In this context, engines and gearboxes as the main components are of twofold customer relevance: First, the complete vehicle performance is mainly determined by the characteristics of the engine (e.g. torque vs. engine speed and response) and the gearbox (e.g. transmission ratio and efficiency), which makes them decisive elements for the vehicle's agility; secondly, both the type of the engine and the type of the gearbox are important customer relevant features, independently from their role in the complete vehicle. Many customers e.g. emotionally prefer a V8 engine over a 6 inline or a gasoline engine over a diesel even though both might lead to a similar or even better vehicle performance. A good example for an engine that is partly requested by customers just for itself is the Chrysler HEMI. The HEMI engine is more or less marketed as a separate product, with its own website and even branded accessories. The HEMI is available in three different brands of the Chrysler Group (Dodge, Chrysler, Jeep), and each car that is equipped with a HEMI bears a logo on the side or trunk. "HEMI inside" can be a direct customer requirement [20].

   Also the choice of gearbox is usually more a matter of personal preference than of optimum performance. Small and economy cars usually feature *manual transmissions* as a basic configuration because they are less expensive and more efficient than automatic transmissions. An automatic transmission might be available as an option, but its comparatively low efficiency can significantly diminish the performance of a small engine. While a manual transmission allows the driver to directly control the power transmitted from the engine to the wheels, an automatic transmission relieves the driver from all coupling and gear changing activities. Thus, sports cars, off-road vehicles and trucks are often equipped with manual transmissions, while comfort-oriented sedans, wagons or SUVs tend to be equipped with automatic transmissions. Different markets also show different preferences: While e.g. the overwhelming majority of U.S. customers prefer auto transmissions (almost 90%), European customers chose more manual gearboxes (80% in the whole market, 60% in the premium market) [4].

   The maximum friction force that can be transmitted by the contact patch between the driving wheels and the road surface limits maximum (acceleration) and minimum (deceleration) vehicle traction. It is mainly determined by the *coefficient*

---

[19] Maximally accelerating the vehicle after traffic or starter lights have switched to green.

of friction (COF) between tire and road surface in the contact patch of the driving wheels [21]. To optimize traction, vehicles may be equipped with four-wheel drive. Four-wheel drive improves traction – especially on slippery ground – and hence is usually chosen for off-road vehicles, but also for improved acceleration, agility, and stability of highly dynamic cars [22].

Customer requirements concerning the braking system include *brake performance* (minimum braking distance) and *sovereignty* (effortless deceleration, stability of the vehicle during the deceleration, minimum body movement during deceleration).

Stability is the tendency of a moving car to stay in a straight line – even when excited by disturbances such as uneven road surface or side wind, especially when accelerating or decelerating at the same time. Lack of stability can be physically felt by the driver e.g. when excessive corrective steering movement is required during an emergency stop.

7.6.1.2 Lateral Dynamics

Lateral dynamic behavior is typically the "fun factor" in vehicle dynamics - and is the hardest part to describe objectively: Steering is perceived by the driver e.g. through subjective feelings such as *directness, talkativeness, informativeness, grip, stability*, or security [23]. Exact control over the vehicle movements up to high rates of lateral acceleration is what sporty drivers are longing for on curvy roads. "It drives like it's on rails" is a typical expression that describes a customer's positive experience of lateral dynamics. A mor technical description is that the transformation of the driver's steering from straight-ahead travel to maximum lateral acceleration – within physical limits – must be spontaneous, predictable, easily controllable and stable, without adverse effects induced by longitudinal excitation (acceleration, deceleration, load alteration), road conditions or climatic conditions.

A vehicle spec that measures lateral dynamics is the maximum possible lateral acceleration of the vehicle. Very few customers actually drive sporty vehicles at the lateral acceleration that was used as the basis for conceptual design. Figure 7.43 compares the lateral load spectrum of a typical customer driving 300,000 km on regular roads (blue line) with that of a professional driver driving the same car 10,000 km on a race track (red line).

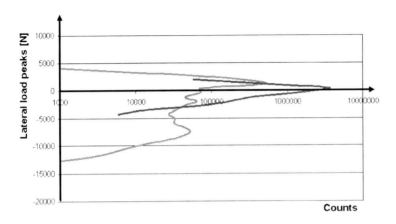

Fig. 7.43 Load spectra for normal and dynamic Driving (Source: BMW)

7.6.1.3 Vertical Dynamics

In terms of agility, vertical dynamics are an enabling characteristic of a vehicle: By ensuring optimum tire-road-contact in every driving situation, vertical dynamics define the physical limits for longitudinal and lateral dynamics. Customer requirements in terms of vertical dynamics are predominantly related to riding comfort and hence are discussed in Sect. 7.4.1.1.

7.6.2 Component and System Design

Design for agility means optimizing conflicting targets such as stability, acceleration/deceleration, traction and comfort. Two subsystems determine a vehicle's agility: Powertrain and chassis.

7.6.2.1 Powertrain

The term *powertrain* denotes the group of components that generate propulsion power and deliver it to the road surface: Engine, transmission, drive shafts, differentials, and drive wheels. The most important design levers for the vehicle's acceleration and traction are the technical parameters of the engine, transmission, and drive wheel tires.

Engine

Different engine types create different driving characteristics. The basic parameters by which reciprocating piston engines for vehicles are differentiated are the fuel type and the respective combustion process (gasoline/diesel), the cylinder arrangement (in-line, V, boxer), and the number of cylinders (3, 4, 6, 8, 10, 12). Additional criteria that stipulate the engine's performance and dynamic behavior are displaced volume, fuel injection technology, air intake/charging technology, and exhaust technology. In Fig. 7.44, different engine configurations (gasoline/diesel/H²; naturally aspirated/charged) are compared in terms of power and torque.

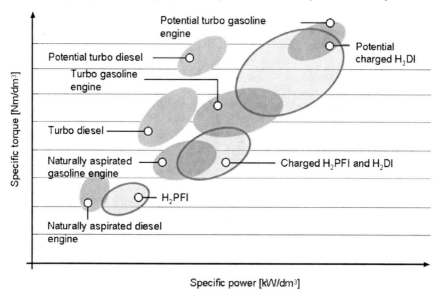

Fig. 7.44 Comparison of engine types (Source: BMW)

An effective – and efficient – method to increase an engine's performance is supercharging. The charger (a compressor in the engine's air intake system that increases the gas volume taken into the cylinder) is actuated through a belt drive to the camshaft (*mechanical charger*) or through a turbine in the engine exhaust (*turbo charger*). The mechanical charger with the required belt drive is usually hard to package to an existing base engine. In any case, a supercharger needs a bypass valve and a charge air cooler as additional components that have to be added to the engine bay. Figure 7.45 compares the effects of mechanical charging and turbo charging on the performance of a 4 cylinder gasoline engine.

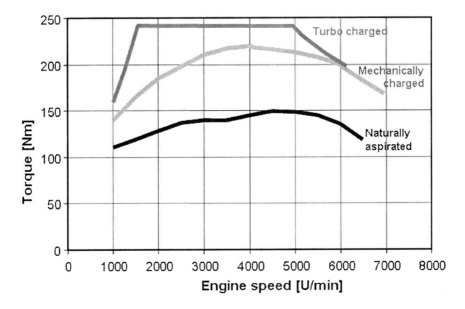

Fig. 7.45 Torque increase by mechanical charging and turbo charging (Source: BMW)

As the additional rotational masses of chargers and turbines have an adversarial effect on the engine response, sporty engines use two small chargers/turbines instead of one big one in order to reduce the total moment of inertia (*bi-turbo*).

Transmission

Just as the engine, the gearbox too must suit the complete vehicle's characteristics and the driver's personal preferences. Prevalent variants of automatic transmissions include:

- Manual transmission: With clutch and floor or column mounted shifter
- Semi-manual transmission: Manual transmission without a clutch pedal; the clutch is engaged or disengaged by a servo controlled electronic transmission control system, e.g. BMW's *SMG* or Fiat's *Selespeed*
- Automatic transmission with torque converter: Conventional automatic transmission with a hydraulic torque converter instead of a clutch, e.g. the ZF 5HP or 6HP family
- Continuously variable transmission (CVT): A continuously variable gear ratio allows the most efficient engine speed to be maintained at almost all vehicle speeds, e.g. e.g. DAF's *variomatic* or AUDI's *multitronic*
- Direct-shift transmission: Two combined manual gearboxes that allow fast and power-loss free gear shifting, e.g. Ford's *PowerShift* or AUDI's *S-Tronic*

Automatic gearboxes follow a shift strategy that derives optimum shift points from input data such as throttle, engine speed, vehicle speed, lateral acceleration, longitudinal acceleration, brake signal and engine temperature [4].

A technical figure of merit that quantifies the sportiness of a gearbox is the minimum gearshift time, which is the time during which the power flow is actually interrupted. In terms of sportiness or performance, direct-shift transmissions today are the superior concept [24].

In the future, transmissions will be individually paired with engines in order to further optimize the complete vehicle's performance [24].

Four-wheel Drive

Compared to a standard two-wheel drive, four-wheel drive systems require an additional transfer case to distribute the engine torque between the front and rear axles, an additional prop shaft, an additional differential and additional drive shafts.

In off-road driving the purpose of a four-wheel drive is to optimally transmit the available torque to all wheels, which usually requires locked shafts (front/rear and left/right). In dynamic cars, where four-wheel drives are used for optimum acceleration and handling, power is precisely distributed between the four wheels by an electronic control system according to the road conditions [4]. Figure 7.46 depicts the components of BMW's xDrive, a four-wheel drive system that detects change in the road surface (for example, while driving in wintry conditions) and instantly redistributes power to counter any under- or oversteering.

Fig. 7.46 xDrive four-wheel drive system (Source: BMW)

With the increasing share of four-wheel drive systems in passenger cars, the future focus in development will be on agility and driving behavior rather than on off-road capabilities, on fuel efficiency (especially with hybrid systems), and on costs [24].

7.6.2.2 Chassis Systems

Together with the basic parameters given by the whole vehicle concept, such as track width, wheel base, center of gravity, weight, axle load distribution, sprung/unsprung masses, or drag coefficient, the chassis components and subsystems (axles, spring-damper-system, wheels and tires, brake system, steering system, and body) determine the overall dynamic behavior (handling and riding comfort) of the complete vehicle [25].

Major conceptual conflicts which have to be resolved during the concept phase are:

- Maximizing wheel width with given engine geometry and vehicle width
- Maximizing spring compression with given hood height
- Optimizing vibrational and acoustics comfort while utilizing new and more complex engine technologies (e.g. hybrid systems, electrical engines ...) and tire concepts (e.g. run-flat tires)
- Optimizing brake performance in spite of package and weight restrictions
- Increasing the usable range of lateral dynamics in consideration of the cost-benefit ratio

A comprehensive description of chassis system development can be found e.g. in [26], [27], and [28].

Suspension Systems

Axles determine track width, wheel base, toe and camber, and incorporate the kinematics of the wheels. Together with the visco-elastic behavior of their rubber bearings and spring-damper elements, these are the crucial parameters of a suspension system. The choice of suspension system must also consider whether the wheels on an axle are driven or undriven and/or steered or unsteered. Suspension systems can be classified as *dependent suspension* systems, in which movements of the left and right wheel are mechanically coupled, and *independent suspension* systems, in which the travel of left and right wheel is autonomous [29]. As dependent systems are usually cheaper and more rigid, they are used in basic vehicles and the heavy-duty segment. Independent systems are chosen when higher requirements in terms of ride comfort and agility must be met. As an example, Fig. 7.47 shows the aluminum 4-link integral rear suspension as used in the 2009 BMW 7 Series.

  As the part of the chassis that reaches farthest up in the vehicle body, spring damper systems are geometrically critical for the complete vehicle concept. For the front axle, the required length of the struts limits the minimum height of the hood, and their diameter and inclination is part of the measurement chain that determines the whole vehicle width (engine, engine beam, strut, wheel house, wheel envelope). For the rear axle, diameter and length of as well as clearance between the struts determine the maximum load space width and thus the practicality – especially for wagons.

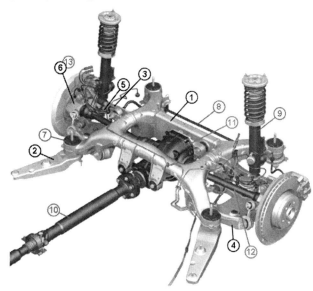

Kinematics:
1   Rear axle carrier
2   Thrust strut
3   Track rod
4   A-arm
5   Transverse
     control arm
6   Wheel carrier

Suspension:
7   Rubber mount
8   Stabilizer bar
9   Spring damper

Propulsion:
10   Propulsion shaft
11   Differential
12   Drive shaft
13   Wheel bearing

Fig. 7.47 Aluminum integral V-rear axle (Source: BMW)

Tires and Wheels

The COF in the contact patch between tire and road is usually the limiting factor for acceleration and deceleration. Aside from the road conditions, the main parameters are:

- Visco-elastic properties of the tire structure and material (as a function of tire temperature and air pressure)
- The tire's macro geometry as a function of its basis geometry (outer diameter, inner diameter, and width) and the lateral, vertical, and longitudinal forces applied to it
- The tire's micro geometry (profile) as a function of wear

Tire specifications – geometry and material – are displayed on the outer side of the tire. Figure 7.48 shows the generic specification of a tire in the U.S.:

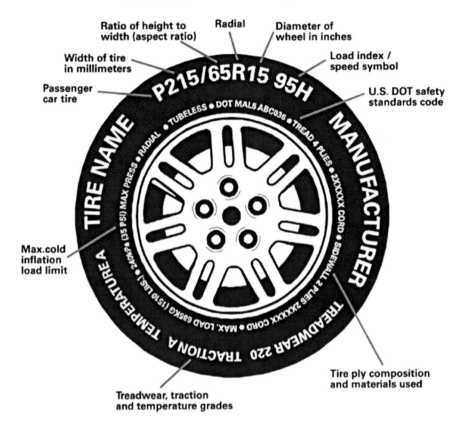

Fig. 7.48 Tire parameters (Source: Internet)

The length-to-width ratio of the contact patch affects the vehicles steering and cornering behavior. For sporty cars, wide and large diamtere tires on big rims are preferred because of their high COF and low rolling resistance on even roads. Off-road vehicles however are usually equipped with tall and narrow tires with a deep profile that ensures optimum friction on rough terrain.

The possible size of the wheels is limited by the dimensions and the y-distance of the wheel houses. The wheel envelope, that is the maximum geometric excursion of a wheel under all driving conditions, as a major input for the complete vehicle concept has already been discussed in Sect. 4.3. As it also limits the maximum diameter of the brake disks, the size of the rims is also a conceptual boundary condition for the brake system.

The biggest share of a vehicle's unsprung masses comes from its wheels. Hence, improved vehicle dynamics can be achieved by usage of – relatively expensive – light weight wheels.

Brake Systems

From the possible three functions a vehicle brake system has to fulfill,[20] controlled and repeatable deceleration is the one that contributes to the vehicle's dynamics. Together with the steering system and tires, the brake system is the sub-system that contributes most to a vehicle's active safety and thus has to meet the highest requirements in terms of reliability. The effectiveness of any brake system is limited by the maximum (negative) traction that can be transmitted by the tire-road contact patch [29]. Other constraints for a service brake system's effectiveness are [4]:

- Actuation force gradient
- Maximum actuation force
- Delay between pedal actuation and wheel deceleration
- Distribution of brake force to the wheels
- Slip control

Today, passenger car service brake systems generally consist of the following components (see Fig. 7.49):

- Brake pedal (activates the dual master brake cylinder)
- Brake booster (amplifies the force that operates the dual master brake cylinder)
- Dual master brake cylinder (creates hydraulic pressure at two pressure outlets)
- Brake hoses and pipes (transmit the pressure to the wheel brakes)
- Brake fluid (transmits the hydraulic pressure)
- Wheel brakes (disk or drum brakes that decelerate the wheels)
- Brake pressure valve (controls the distribution of pressure between front and rear wheels)
- Electronic control systems to support optimum braking in every driving situation (see below)

The higher the required amplification, the bigger the required diameter of the brake booster – which makes it together with its *left-hand drive/right-hand drive* (LHD/RHD) *dependency* an important factor in the engine bay package. In particular, the air ducts required to cool the front brakes are critical for the front-end package and design. The elasticity of the brake hoses is the main factor that determines the system response time. The maximum size of the brake disks is limited by the wheel size. All parts of the wheel brakes (disks and calipers or drums and shoes) add to the unsprung masses and thus are relevant for vertical dynamics.

Together with the electronic chassis control systems discussed below, a future concept that bears huge potentials for alternative package concepts is *brake by wire*. Electrically operated wheel brakes eliminate the retarding visco-elastic and hydrodynamic effects from fluid filled brake lines and the package restrictions

---

[20] Deceleration of the vehicle (service brake system), ensuring constant speed when driving downhill (supplemental brake system), hold the vehicle stationary (parking brake).

caused by the master brake cylinder and brake booster. A potential solution for electrically actuated wheel brakes are wedge brakes, a concept in which the self-hemming effect of wedge-shaped brake pads is electronically controlled. The critical aspect for both concepts however is the reliability of the electronic control system that, in the case of malfunction, could lead to fatal risks caused by brake failure or immediate wheel blockage.

1 Wheel brakes, wheel-speed sensors,
  brake-pad wear sensors
2 Brake pedal cluster (brake pedal, brake
  servo unit, tandem master cylinder)
3 Brake fluid reservoir
4 ABS/DSC control unit and
  hydraulic modulator
5 Brake hoses / pipes
6 Hand brake lever
7 Hand brake
  bowden cable

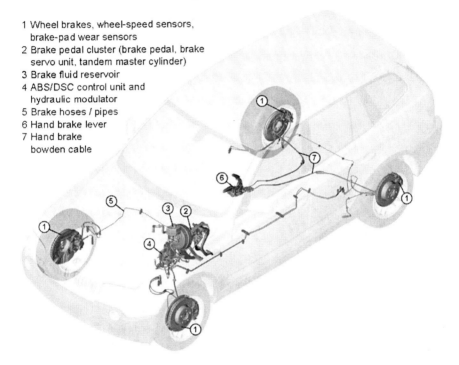

Fig. 7.49 Service brake and parking brake system (Source: Internet)

Steering Systems

The vast majority of passenger cars today use power assisted rack and pinion steering systems, where the steering wheel turns the steering column with the pinion gear attached, the pinion moves the rack (a linear gear meshing with the pinion) from side to side. The rack is linked to the track rods which move the steered wheels via the steering arms. Included in the steering column is a torsion bar that measures the torque applied to the steering wheel and operates a hydraulic valve (hydraulic power steering) or an ECU (electronic power steering) that triggers an hydraulic or electric actuator attached to the steering rack or column (see Fig. 7.50).

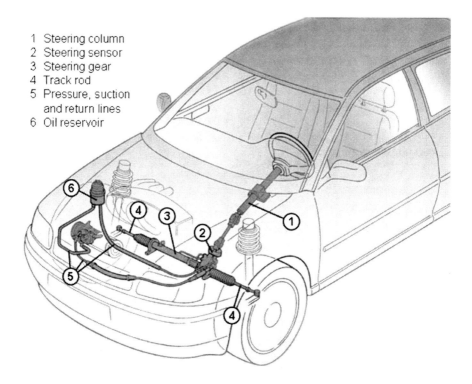

1 Steering column
2 Steering sensor
3 Steering gear
4 Track rod
5 Pressure, suction
  and return lines
6 Oil reservoir

Fig. 7.50 Components of a hydraulic power steering system (Source: ZF Lenksysteme)

Conventional rack and pinion steering systems with a fixed gear ratio have to compromise between a very responsive steering that is pleasant when driving on curvy roads at moderate speed (or especially when parking) and a less responsive steering that gives the car directional stability at higher speeds. A steering concept that dissipates these competing requirements is e.g. BMW's AFS, where a planetary gear integrated into the steering column steplessly alters the gear ratios, thus creating an increased steering angle at low speed or a decreased steering angle at high speed. A permanent mechanical connection between the steering wheel and wheels remains.

A concept for the future that would allow further flexibility in terms of tuning a vehicle's steering behavior is *steer-by-wire*, where the steering forces are solely generated by electrical actuators – without any mechanical connection between the steering wheel and the steered wheels. A feedback actuator simulates a realistic steering feel for the driver [30]. Apart from allowing steering features such as a steering ratio that is continuously adapted to driving conditions or automated parking assistance, the ability to eliminate the steering column opens vast potentials and improvements through:

- LHD/RHD-independency of the engine bay
- Improved crash performance
- Simplified assembly process

Similar to brake-by-wire, safety is currently the biggest concern in the application of steer-by-wire systems. The lack of a mechanical connection between the steering wheel and the steered wheels poses extremely high demands on the system reliability, which is achieved through complete system redundancy (see Fig. 7.51).

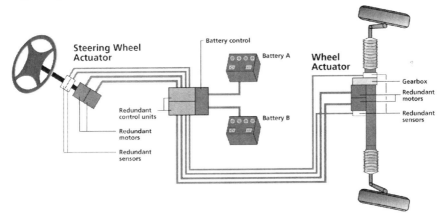

Fig. 7.51 Steer-by-wire system (Source: ZF Lenksysteme)

Electronic Chassis Control Systems

Since traditional passive design has been almost driven to its limits, active electronic control systems are today in the foreground of research and development. Electronic chassis control systems can be classified in *wheel slip control* and *yaw moment control* (GMR) [4].

Wheel slip, the relative motion between a tire and the road surface, decreases the COF in the contact patch and thus the possible traction. An electronic wheel slip control system constantly computes for the slip at each wheel as the difference between the rotational speed of this wheel and the average rotational speed of all wheels. Widely used wheel slip control systems include:

- Anti-lock braking systems (ABS), which maintain the vehicle's steerability and stability during deceleration by rapidly reducing the brake force when the wheel slip passes a certain limit.
- Brake force distribution systems, which distribute the available braking force to the four wheels in a way that optimizes the vehicle's braking performance. *Electronic brake force distribution* (EBD) systems distribute the brake force

between front and rear axle; *cornering brake control* (CBC) systems do this between right and left side.

- *Traction control systems* (TCS) maintain the maximum possible traction during acceleration by reducing the torque transmitted to drive wheels when their slip reaches a certain limit. Torque reduction can be realized either by reducing the engine output or by braking the appropriate wheel.[21]

Yaw moment control improves a vehicle's lateral stability through counteracting possible under- or oversteering by braking the inside rear wheel or the outside front wheel respectively, as soon as the continuous measurement of lateral acceleration, yaw rate, and steering angle indicates a potentially hazardous dynamic state of the vehicle.

Damper control systems use shock absorbers with adjustable visco-elastic characteristics to optimize a vehicle's vertical dynamics both in terms of handling and riding comfort – with regards to driving style and road conditions.[22]

Electronic stability control (ESC) systems[23] integrate wheel slip control, yaw moment control, and damper control systems to comprehensively optimize a vehicle's stability and maintain control during all dynamic driving conditions. An ESC system identifies emerging critical driving situations (such as skidding, rolling, or wheel blocking) and reacts through targeted intervention in engine, brakes, and dampers.

Brake assistant (BA) systems recognize an intended emergency brake and immediately increase brake pressure in order to minimize the braking distance.

In all of these systems, sensors (e.g. for wheel speed, yaw, transversal and rotational acceleration etc.) detect the actual status of the vehicle, an ECU with the appropriate software identifies problematic conditions and determines counteractive measures which are then executed by actuators. An *integrated chassis control* concept optimizes complete vehicle characteristics rather than individual functions, making the best use of all available sensors and actuators (see Fig. 7.52).

Chassis control systems must meet maximum safety requirements. The methods of E/E systems engineering that allow immediate response and highest reliability are discussed in Sect. 5.2.

---

[21] Brand-specific systems are e.g. ASR/ASC+T (BMW) or ASR/ETS (Mercedes).

[22] Brand-specific systems are e.g. BMW's EDC (Electronic Damper Control) or Jaguar's CATS (Computer Active Technology Suspension).

[23] Brand specific trade names for ESC systems are e.g. DSC (Dynamic Stability Control), ESP (Electronic Stability Program), or VDC (Vehicle Dynamic Control).

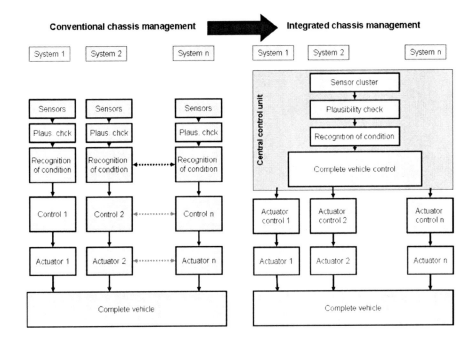

Fig. 7.52 Integrated chassis management (Source: BMW)

Body

In today's passenger cars, monocoque bodies have replaced ladder frames as the mounting surface for suspension, steering and brake components. Body type and geometry determine its longitudinal and torsional stiffness, major parameters of a vehicle's dynamic behavior [31]. They establish the rigidity of the link between front axle and rear axle. Figure 7.53 compares body variants in terms of their longitudinal and torsional stiffness, taking the derivatives of the BMW 3 Series as an example.

Elements expanding in y-z (e.g. windshield, rear window glass, dash board cross member) contribute to the torsional stiffness, which can be increased by strut braces in the engine bay, reinforcements in y-z in the cabin (e.g. triangular enforcements between b-pillar and floor), or e.g. self-locking tail gates in wagons and SUVs. As an example, Fig. 7.54 shows stiffness reinforcements to the body of a BMW 3 Series coupe.

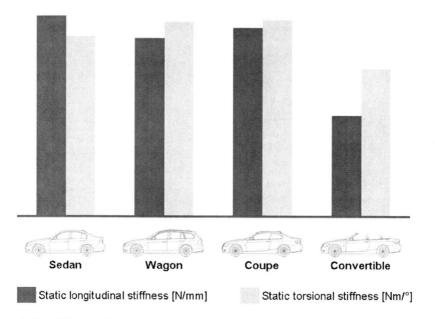

Sedan   **Wagon**   **Coupe**   **Convertible**

■ Static longitudinal stiffness [N/mm]   Static torsional stiffness [Nm/°]

Static stiffness values measured over domes, with glazing, frontend,
front axle carrier, supporting tube, without rear axle carrier and sunroof

Fig. 7.53 Torsional and longitudinal stiffness of body variants (Source: BMW)

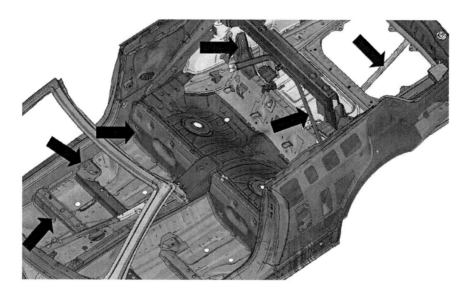

Fig. 7.54 Stiffness reinforcement in a convertible's body structure (Source: BMW)

7.6.3 System Integration and Validation

7.6.3.1 Performance

Evaluation of the performance characteristics of an engine starts with early simulation e.g. of the combustion process or the thermodynamic behavior of intake and exhaust systems. Initial SIL and HIL testing (compare Sect. 5.2.5.3) is then followed by hardware component tests until the whole engine or the whole power-train is tested in a test rig (see Fig. 7.55). In HIL testing, hardware that is physically absent (e.g. tires, vehicle, road or driver) is simulated in real-time and thus included in the testing.

Fig. 7.55 Powertrain test rig (Source: BMW)

Complete vehicle validation then starts with the first prototype build group and is first carried through on roller dynamometers, then on closed test tracks, and eventually on public roads (see Fig. 7.56).

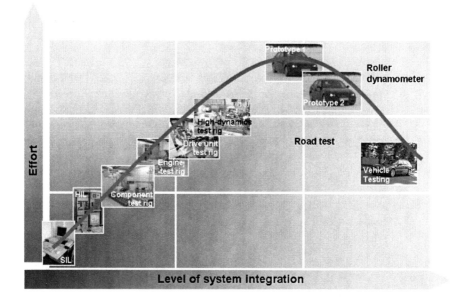

Fig. 7.56 Stages of powertrain validation (Source: BMW)

In a conventional engine test rig, stationary engine behavior for specific operating points is observed by applying a constant counter torque by means of a braking system. A *dynamic engine test rig* uses an electric load machine to simulate realistic driving cycles with fast changes of torque or engine speed. The latest development however are *high-dynamic engine test rigs* which realistically simulate complete vehicle dynamic behavior in real time (e.g. rotary oscillations superimposed to the crankshaft rotation), so that test results which formerly have required prototype testing now are available with only the engine available in hardware. High-dynamic test bays dramatically reduce the overall testing effort and lead to an early product maturity. The comparison of the engine data received in tip-in mode from a real road test and a high-dynamic rig test shows the accuracy of the simulation (see Fig. 7.57).

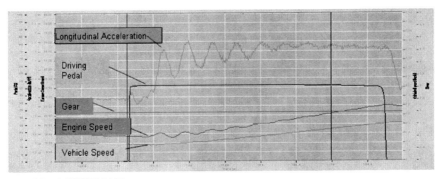

Road test (BMW 318i)

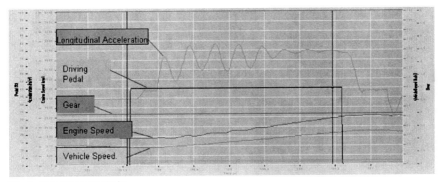

High-dynamic test rig

Fig. 7.57 Road test and high-dynamic test rig in tip-in mode (Source: BMW)

7.6.3.2 Handling

At the beginning of a vehicle project, the chassis concept is composed from pre-developed building blocks and evaluated by simulation. The complete vehicle is modeled as a *multi-body system* (MBS) that is enhanced by elastic elements (with defined spring rates and damping characteristics) to evaluate the elasto-kinematic characteristics of suspension and steering (see Fig. 3.5).

Today, conceptual simulation of complete vehicle dynamics is mostly used to evaluate the vehicle handling during standardized open loop driving maneuvers such as circular driving at constant speed, step steering, or sinusoidal input steering, where the impact of the driver's individual habits on the vehicle's reaction is eliminated to the greatest possible extent [25].

While off-line simulation is able to replicate mechanical behavior of components, the overall perception of vehicle dynamics is highly subjective and needs an experienced test driver as a sensor. Driving simulators create the possibility to have a real driver experience and evaluate the agility of a vehicle whose drivetrain and suspension only exist virtually: A car body equipped with a real driver environment (seat, pedals, steering wheel etc.) is mounted on a hydraulic hexapod platform[24] and thus can be accelerated in all degrees of freedom (Fig. 7.58 right). As visual feedback is essential for experiencing dynamics, the vehicle's environment is projected on a screen in front of the simulator's cabin (Fig. 7.58 left).

Fig. 7.58 Driving simulator (Source: BMW)

To a certain extent, the high speed control systems of a driving simulator allow a test driver to simulate driving maneuvers in real time and experience both handling and comfort characteristics. Extreme maneuvers that would exceed the kinematics and dynamics capacity of the hexapod platform can not be simulated in real time; they can however be pre-calculated from given input data and then replayed by the driving simulator. A big advantage of the driving simulator, even compared to real vehicle testing, is the possibility to quickly switch between different concepts to directly compare them – without having to stop and switch test cars in between.

When evaluating agility with a real car on real roads, two different targets are most important: The subjective evaluation of the vehicle's characteristics as perceived by the driver, and the measurement and analysis of objective dynamic data. Figure 7.59 shows a test car that is equipped with the corresponding measuring equipment.

---

[24] A platform that is actuated by an octahedral assembly of six interlinked hydraulic struts.

Fig. 7.59 Test vehicle with dynamics measuring equipment (Source: BMW)

7.7 Passive Safety

7.7.1 Legal and Customer Requirements

7.7.1.1 Active and Passive Safety

Safety as a complete vehicle characteristic depicts both the vehicle's ability to reduce the risk of being involved in an accident (*active safety*) and – if an accident can not be avoided – to avoid or mitigate injuries or damage to passengers and the car (*passive safety*). In order to make a vehicle as safe as possible, an integrated safety strategy considering all parts of the logical chain that may lead to an accident has to be followed during development. The phases of this process and their requirements in terms of safety are illustrated in Fig. 7.60.

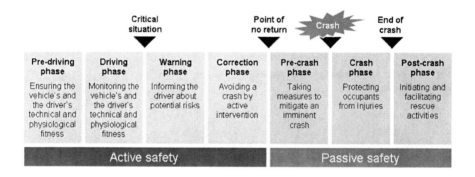

Fig. 7.60 Phases of crash avoidance and mitigation (Source: BMW)

Customer requirements for active safety are – to the greatest extent – incorporated in other complete vehicle characteristics which have already been discussed:

- Design, e.g. visibility through shape and color, driver all-round vision through body design (see Sect. 7.3)
- Cabin comfort, e.g. driver alertness through noise level, thermal comfort, or vertical dynamics (see Sect. 7.4)
- Infotainment, e.g. reduced driver distraction through ergonomic displays and controls, driver alertness through traffic information and advanced support systems (see Sect. 7.5)
- Agility, e.g. short braking distance through brake assistant, or ability to avoid an obstacle through ABS and ESC (see Sect. 7.6)

7.7.1.2 Assessing Vehicle Safety

In contrast to almost all other complete vehicle characteristics, passive safety can not be personally evaluated by the customer prior to purchase, and even after purchase it can only be experienced in the rare and unwanted case of an accident. Hence, customers seek proof – or evidence – of how well they and their passengers would be protected from injuries if involved in an accident. This is achieved in two ways: By listing the safety features (such as airbags or seatbelt pretensioners etc.) the car is equipped with, and through the results of standardized crash tests, which allow evaluation and rating of a vehicle's passive safety performance. Derived from the analysis of actual road accidents – as e.g. done by the *European Transport Safety Council* (ETSC) for accidents within the European Union in 2001 [32] – the main types of crash tests are:

- Frontal impact (rigid or deformable barrier, full width or offset, straight or oblique)
- Side impact (plain barrier or pole)
- Rear impact (rigid or deformable barrier, full width or offset, straight or oblique)
- Rollover

Additional emphasis is placed on:

- Child protection
- Pedestrian protection

Conflicting goals in the specification of crash-tests are the wish to simulate real-world accident situations on the one hand and the need to get reproducible values that allow distinct vehicle evaluation on the other hand. However, real-world accidents typically involve two or more cars, and the effects the crash has on one car depends largely on the behavior of the others. Thus, the results are not suitable as an absolute rating for the car under investigation.[25] In order to obtain an objective passive safety rating of distinct vehicles, standardized barriers substitute for the second car in most crash tests. Tests must be precisely specified so that reproducible results are delivered (see below).

7.7.1.3 Mandatory Evaluation

In all markets, the legal requirements which must be met to approve a car for registration include targets for passive safety performance that will guarantee a minimum statutory standard of safety for new cars. Mandatory safety equipment and crash tests relevant for registration are specified e.g. by FMVSS in the U.S. or

---

[25] Some manufacturers however perform crash tests between two mostly very dissimilar vehicles to evaluate and demonstrate their passive safety under realistic conditions.

by ECE regulations in Europe (see Sect. 7.1.3). For North America and Europe, legally required tests today include:

- U.S./Canada

    – Frontal crash FMVSS/CMVSS 208/301
    – Side crash FMVSS/CMVSS 214/301
    – Rear crash FMVSS/CMVSS 301

- European Union

    – Frontal crash ECE-R94
    – Side crash ECE-R95

In addition to these crash tests, also passing quasi-static tests such as the roof intrusion test according to FMVSS/CMVSS 216 is required for approval.

7.7.1.4 Non-mandatory Evaluation

As many customers want their cars to exceed these minimum requirements and wish to obtain a quantitative rating of a vehicle's safety performance prior to purchase, comprehensive non-mandatory test programs for passive safety have been established by manufacturer independent groups with representatives from governments, automobile groups and insurance companies. In Europe, the *New Car Assessment Programme* (Euro NCAP) awards star ratings to new cars based on their performance in standardized crash tests. Figure 7.61 illustrates the parameters for the Euro NCAP frontal impact crash, Fig. 7.62 for the side impact crash.

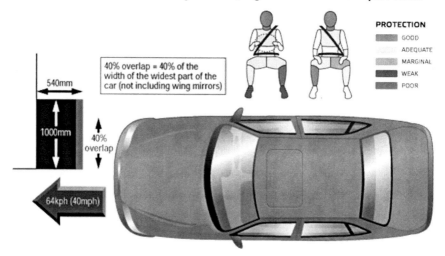

Fig. 7.61 Euro NCAP frontal impact assessment and rating (Source: Euro NCAP [33])

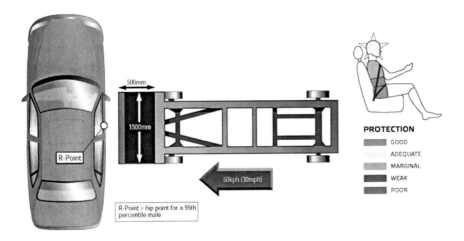

Fig. 7.62 Euro NCAP side impact test assessment and rating (Source: Euro NCAP [33])

In addition to crash tests, Euro NCAP testing includes the assessment of accidents with child and adult pedestrians. As depicted in Fig. 7.63, during theses tests different parts of the vehicle's front body are evaluated in terms of the injury risk they pose to leg, upper leg and head.

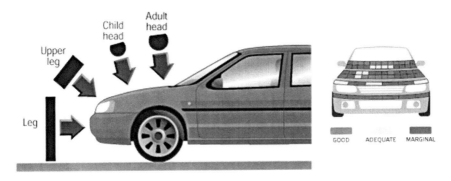

Fig. 7.63 Euro NCAP pedestrian protection assessment and rating (Source: Euro-NCAP [33])

Eventually, the overall Euro NCAP score is calculated as the total of all points a vehicle could achive in 10 different test categories – distributed over 3 actual crash tests (frontal, barrier side, pole side). Table 7.5 lists these test categories and the maximum number of points per category. The maximum overall score is 36 points.

Table 7.5 Euro NCAP scoring table (Source: Euro NCAP [34])

Test	Region of injury	Maximum number of points
Frontal	Head and neck	4
Frontal	Chest	4
Frontal	Knee, femur, pelvis (i.e. left and right femur and knee slider)	4
Frontal	Leg and foot (i.e. left and right lower leg and foot and ankle)	4
Barrier side	Head and neck	4
Barrier side	Chest	4
Barrier side	Knee, femur, pelvis (i.e. left and right femur and knee slider)	4
Barrier side	Leg and foot (i.e. left and right lower leg and foot and ankle)	4
Pole side	Head and neck	2
Pedestrian		2

The final star ratings (5 to 0) then are generated from both the overall scores and the balance between side and front scores (see Table 7.6).

Table 7.6 Euro NCAP star rating conversion table (Source: Euro NCAP [34])

Number of stars	Required min. overall score [points]	Required min. single test score (frontal / side) [points]
5	33	13
4	25	9
3	17	5
2	9	2
1	1	
0	0	

NCAP assessments also exist in the U.S., Japan and other markets. In the U.S., different crash assessments for new cars are carried out by NHTSA (so-called US NCAP) and the *Insurance Institute for Highway Safety* (IIHS), a U.S. non-profit organization funded by auto insurers. While the newer IIHS crash tests are similar (though not identical) to the Euro NCAP, US NCAP tests differ in overall setting, barrier type, offset, dummies and rating. In contrast to the Euro NCAP the US NCAP rating represents the likelihood of a severe injury, with 1 star representing 25% and 5 stars representing 5% likelihood. This divergence can lead to identical cars obtaining different ratings in Europe and the U.S.. The goal of the ongoing attempts to globally harmonize safety requirements (see Sect. 7.1) is the *World*

NCAP with internationally established test procedures – even though current development of both the Euro NCAP and the US NCAP are tending to diverge rather than converge.

As a manufacturer-independent figure for a vehicle's passive safety, NCAP ratings have become an important factor for the customer's purchase decision of new vehicles all over the world. Car manufacturers incorporate the respective test requirements into their designs, and propagate positive ratings as a marketing message.

On the other side, bad ratings have an immediate impact on marketing and sales as well: When the General German Automobile Association (ADAC) published the devastating results the then newly introduced Brilliance BS6 achieved in a Euro NCAP frontal impact crash test in June 2007 (see Fig. 7.64), Brilliance had to realign their market introduction strategy for Europe.

Fig. 7.64 Vehicle with collapsed cabin after Euro NCAP frontal crash (Source: ADAC)

NCAP ratings are however not always an absolute indicator for a vehicle's safety. A five star rating e.g. requires availability of a seat belt reminder. For a cautious driver however who always wears a seat belt, a car without a seat belt reminder (and hence with a maximum of four stars) is not less safe.

7.7.1.5 Vehicle-independent Factors Determining Safety

Even though mitigation of injuries and casualties are of the highest importance in passive safety activities, the minimization of the damage an accident does to the car itself is a another customer requirement. An example is the *bumper standard* FMVSS 49 CFR Part 581 which requires that frontal or rear collisions under 5 km/h may not lead to any damage at all. As this aspect is more closely related to the cost of ownership than to passive safety, it is discussed in Sect. 7.2.1.2. For the safety of motor traffic in general, the technical features and capabilities of the vehicles (the main focus of this section) are only one of several criteria. Other factors that determine how safe it is to drive a car are e.g.:

- Traffic regulations, e.g. speed limits, passing regulations, allowed level of blood-alcohol etc.
- Road infrastructure, e.g. road surface, road geometry, signposting, illumination, protective devices for pillars, guard rails etc.
- Drivers' maturity: Minimum driving age, requirements to obtain a driver's license, enforcement of safety-critical traffic violations etc.

7.7.1.6 Injury Evaluation

Quantitative evaluation of passive safety requires a method to objectively measure and compare the severity of the occupants' injuries. Within the worldwide automotive industry, the *Abbreviated Injury Scale* (AIS) issued by the *American Association of Automotive Medicine* (AAAM) has become the standard method for this purpose. According to AIS98, injuries are classified by location and severity. The location is indicated by the 6 digit anatomical localizer (AIS98-ID). Table 7.7 shows the first level (indicated by the first digit) of the AIS98-ID.

Table 7.7 Injury location according to AIS98 (Source: AAAM)

AIS98-ID (first level)	Body region	Number of AIS98-IDs
1	Head	236
2	Face	87
3	Neck	80
4	Thorax	175
5	Abdomen	233
6	Spine	208
7	Upper extremity	125
8	Lower extremity	164
9	External and other trauma	33

The severity of an injury is quantified by the 1 digit AIS code (see Table 7.8). Safety requirements are usually given by a MAIS value which denotes the maximum AIS value.

Table 7.8 Injury severity rating acc. to AIS98 (Source: AAAM)

AIS98 Code	Severity
1	Minor
2	Moderate
3	Serious
4	Severe
5	Critical
6	Maximum
9	NFS (not further specified)

An enhancement to the AIS scale is the *Injury Cost Scale* (ICS) that allows an estimate of the overall costs associated with certain injuries [35]. Decisions on legal measures to improve traffic safety are calculated based on the potential savings in the costs of injuries prevented in a governmental cost/benefit analysis [36].

7.7.2 Component and System Design

The overall degree of passive safety a vehicle offers to its occupants – and other traffic participants – depends on the following factors:

- Safety-related properties of vehicle components, especially the body but also seats, steering wheel, trim parts etc.
- Availability and quality of safety features such as seat belts, air bags, seat belt pre-tensioners, roll-over bars etc.

7.7.2.1 Vehicle Components

Body

Without any doubt, the vehicle body's crashworthiness is the most important factor in passive safety. Here, its twofold task is to allow planned structural deformation and the associated energy absorption (especially in frontal or rear impacts) and at the same time to provide the rigidity that prevents or diminishes the intrusion of another body into the cabin (especially in side impacts and rollover) [37].

In a front-impact, controlled deformation of the crumple zone leads to controlled deceleration of the restrained occupants which in turn limits the forces occupants are exposed to and thus minimizes injury. During this deformation process, the major part of the energy is absorbed by the longitudinal members, which therefore have to be designed in a way that their collapsing behavior is predictable. Figure 7.65 shows longitudinal members before and after deformation. On a smaller scale, the same process applies to the rear vehicle structure in case of a rear impact.

Fig. 7.65 Controlled collapse of longitudinal members
(Source: BMW)

To obtain the required rigidity for side-impacts or rollovers, critical parts of the body structure – such as the B-pillar or side door impact bars – are made from ultra high strength steel. Figure 7.66 shows an example of this on the BMW 1 series.

Due to the restricted space, absorption of kinetic energy in side impacts is only achieved through deformation of the side frame and door panel as well as through the impulse pushing the vehicle away. On the struck side, passengers can be protected from injuries by means of air bags, padding material, or bolsters.

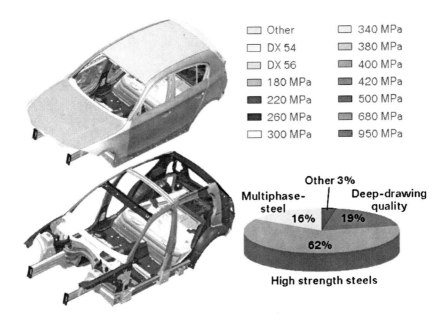

Fig. 7.66 Steel grades in vehicle bodies (Source: BMW)

While the body is the vital component during a crash, it also plays an important role in the post crash phase. Deformation of side frame and doors as well as the behavior of locking systems determine how easy it is to open the vehicle doors after a crash to rescue occupants trapped in the damaged vehicle.

Even minor changes to the body structure can significantly alter its collapsing behavior. As this in turn would require additional costly crash tests for passive safety validation, every effort must be made to keep the body design as stable as possible during series development.

Bumpers

As their name suggests, bumpers do not notably contribute to a vehicle's crashworthiness or occupant protection but prevent or reduce physical damage to the front and rear ends of a vehicle in frontal and rear low-speed collisions – the by far most frequent type of collision.

Legislation requires bumpers to be capable of sustaining low speed frontal and rear crash without any damage to the vehicle structure (compare Sect. 7.7.1.5). Figure 7.67 shows the deformation of the front bumper of a BMW X5 in a low-speed frontal impact. The movable measuring stick mounted on the hood indicates the actual travel distance of the bumper. The goal to minimize damage to the bumper however conflicts with the styling-related goal of offering bumpers painted in the color of the body which makes them much more susceptible to damage from even minor collisions.

Fig. 7.67 Front bumper before, during, and after a low speed frontal collision (Source: BMW)

As the bumper is the part o a car that actually collides with the other party in a frontal or rear crash, its geometry and visco-elastic properties are crucial for the safety of outside traffic participants such as pedestrians (compare Fig. 7.63), cyclists and motorcyclists.

Steering Wheel, Steering Column, Dashboard and Trim Parts

To mitigate potential occupants' head injuries, airbags prevent contact between head and steering wheel, dashboard, or side frame – formerly the typical cause for severe head injuries. Also trim parts which can collide with the occupants' extremities, e.g. the lower side of the dashboard with the knees are covered with padding material.

Another approach to prevent this kind of injury is retraction of the steering column as e.g. realized in the Procon-ten system introduced by AUDI in the 1980s. Procon-ten used steel ropes fixed to the engine that – in the event of a frontal impact – would pull the steering column back and at the same time tighten the seat belts. However – especially at higher speeds, the extremely fast rearward movement of the steering wheel imposed substantial risks of injury to the driver resulting in the system being discontinued in the succeeding generation of cars.

7.7.2.2 Safety Features

Seat Belts

Having been introduced by Saab and Volvo in the late 1950s, seat belts are still the most important safety feature in a car. Most of the advanced safety features such as airbags etc. can offer only limited protection to passengers not wearing a three point seat belt. For this reason, most countries require new vehicles to have all seats equipped with three point seat belts that limit forward motion of the occupants, absorb crash energy and keep occupants from moving through or being ejected from the vehicle. Not wearing a seat belt is still the main reason for fatal injuries in car crashes. Unbelted rear passengers e.g. also increase the risk of death for belted front passengers up to a factor of five [38].

The design of the upper attachments of the front seat three-point seat belts has a direct impact on body and seat design. While the two lower anchor points of three-point seat belts are usually fixed to the seat structure body floor, there are two main concepts for the upper one:

- Mounted to the B-pillar: The force created by decelerating the occupant's body is directly transmitted into the B-pillar. As the backrest is not part of the load transfer path, it can be designed in a simpler and lighter way. In some cars, fitment of the can be optimized by means of an upper attachment that is adjustable along the B-pillar – which in turn weakens the structure of the B-pillar.
- Mounted to the seat's backrest (belt-in seat): Allows a body structure without B-pillar (as desired e.g. in convertibles or coupes), and makes rear passenger entry and exit easier in two-door vehicles. Also, fitment of the belt becomes independent of the seat position. This design requires especially strong rear brackets, seat frame and backrest to compensate for the forces resulting from the wearer's forward movement.

To improve injury prevention and wearing comfort, technical enhancements have been made to seat belts over time:

- *Emergency locking retractors* (ELR) allow free occupant movement while driving, but locks the seat belt when a collision or rollover is detected.
- *Automatic locking retractors* (ALR) offer a ratchet mechanism allowing manual locking of the seat belt to secure a child seat (mandatory in the U.S.).
- *Pre-tensioners* tighten the seat belts instantaneously if an imminent collision has been detected. By means of electric or pyrotechnic actuators, slack is taken out of the belt and occupant travel is limited.
- *Force limiters* allow the seat belt to yield in a controlled way during a collision to lower the peak level of the forces the seat belt exerts to the occupant.

Although wearing seat belts is mandatory in almost all countries today, many drivers and passengers still do not buckle up. Seat belt usage rates for front seat occupants were at 81% in the U.S. in 2006 [39] and at 76% in the European Union in 2003 [36]. Hence, although they do not contribute to a vehicle's passive safety, seat belt reminders reduce the overall risk of injuries by increasing the seat belt use rate.

Airbags

Airbags or – as they are supposed to be used in addition to seat belts – *supplemental restraint systems* (SRS) are inflatable nylon bags that cushion the impact of the head or body parts to rigid interior components in a crash situation. There are four different kinds of airbags.

- Front airbags protect the heads and upper torsos of the front row occupants from hitting the steering wheel or cockpit.
- Side or thorax airbags cushion the torso in side impacts.
- Curtain airbags protect the head and upper body of passengers in side impacts.
- Knee airbags protect the driver's knees in frontal impacts.

Figure 7.68 depicts form and position of the front, side and thorax airbags in a BMW X3.

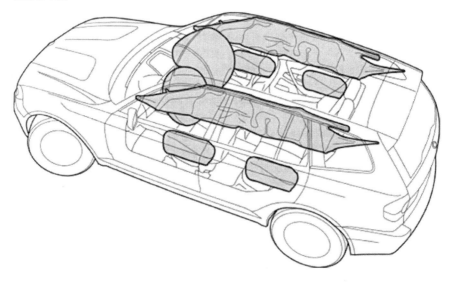

Fig. 7.68 Shape and location of front and side airbags (Source: BMW)

The sequence of events in a frontal collision starts with its onset, the first contact between the vehicle and the obstacle. About 15 to 30 ms after the onset, the vehicle acceleration values detected by the sensor cluster reach or pass the required threshold to indicate the state and type of a crash. Together with other measures like the deployment seat belt pre-tensioners the airbag control unit ignites pyrotechnic gas generators that blow up the relevant airbags within 20 to 30 ms. Injury to the occupants is prevented by controlled deceleration achieved by having the gas vent out of the airbags through small vent holes after contact with the occupant. Deployment and inflation of the airbags as part of the complete sequence of events during a crash is shown in Fig. 7.69. Additional safety measures are unlocking the doors to facilitate rescue, or disconnecting the battery and shutting off the fuel pump to prevent sparks from short circuits igniting a fuel fire.

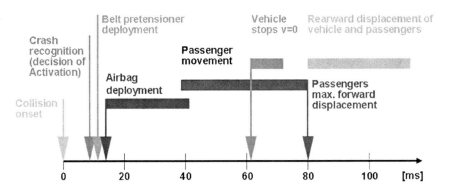

Fig. 7.69 Sequence of events during a crash (Source: BMW)

Post-crash Systems

While passive safety systems mitigate injury during a crash, post-crash systems are there to get help to injured occupants as soon as possible. In case of a crash, systems like OnStar's *Advanced Automatic Collision Notification* (AACN) detect crucial information such as the exact location, type and severity of the crash, number of occupants, airbag deployment etc. by means of distributed sensors in the vehicle. This information is then sent by the vehicle to an emergency call center where it is forwarded to the relevant emergency dispatch center. Early availability of this information enables emergency and medical services to optimally prepare recue and treatment. A similar approach is followed by *eCall*, a project of the European Union that will make rapid automatic assistance available in all member states by 2010.

7.7.3 System Integration and Validation

7.7.3.1 Crash Simulation

As all crash tests include destruction of the respective vehicle or component, testing passive safety through real crash tests is extremely expensive. The costs for an early prototype can easily exceed one million Euros. This is why among all complete vehicle characteristics, passive safety is the one for which the most elaborateed simulation methods are available today. Simulation results are in fact so precise, that in the framework of a vehicle development project plan, hardware testing is expected to prove their results, and not to create substantially different findings. Taking the Euro NCAP Frontal Impact crash of a BMW X5 as an example, Fig. 7.70 compares the results of a virtual and a real crash test.

Fig. 7.70 Comparison of crash simulation and real crash test (Source: BMW)

The main results that come out of the investigation of a virtual car crash are:

- The process of geometric deformation of the vehicle, in which the remaining occupant space (e.g. through collapse of a pillar) and deformation of components that lead to potential injuries (e.g. intrusion of the steering column into the cabin) are of interest
- The deceleration data for the passengers, detected by means of virtual crash dummies
- The effectiveness of passive safety features such as seat belts, airbags, or belt pre-tensioners

Figure 7.71 shows four examples of crash simulation. In the top left picture, the performance of safety features for the front row passenger is evaluated. The top right picture demonstrates the use of a virtual human body model to evaluate injuries in a side crash. In the bottom row, two examples of complete vehicle crash simulations are shown: An early concept simulation of a BMW 7 Series on the left side, and a detailed simulation of a BMW X5 with the occupants and their environment extracted from the car for improved visibility of crash relevant occurrences.

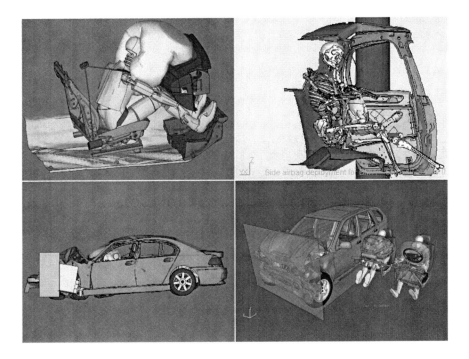

Fig. 7.71 Complete vehicle and component crash simulations (Source: BMW)

Although simulations can replace real crash tests to a certain extent, crash algorithms today still have certain limits that make hardware tests necessary. Vehicle properties that can not be modeled sufficiently include:

- Behavior of non-plastic materials (e.g. glass, fiber-reinforced plastics)
- Geometric tolerances
- Fuel tightness of the fuel system
- Jamming of doors

7.7.3.2 Crash Tests

To confirm that results obtained from simulation resemble real vehicle behavior, crash tests are performed during every hardware prototype build group. If the real test reveals a different picture than the simulation, additional development loops are required causing delay in the time plan and additional costs. To be able to react as quickly as possible in this case, usually one of the first cars of a prototype build group is used for a crash test.

At the end of the PEP, conformity with mandatory and non-mandatory requirements is then proved by performing crash tests with pre-series cars that not only have the product status that will be sold to customers later on but also have been produced by series production processes.

Crash Test Dummies

Naturally, fulfillment of customer requirements in terms of passive safety can not be evaluated by human test drivers.[26] Objective data concerning the injury level caused by standardized crash tests are obtained by means of crash test dummies, *anthropomorphic test devices* (ATD) with a realistic geometry and weight distribution that are equipped with sensors recording forces, accelerations, or deformations of the different parts of their bodies during a collision. Figure 7.72 shows the sensors of a Hybrid III dummy.

According to the prevalent injuries of each crash type, different dummies with different measuring equipment installed are used for frontal impacts, side impacts, or rollovers. In order to produce comparable results, crash dummies are standardized, and each specification of a crash test includes the required type of crash dummies. Today, the prevalent dummy types are:

- *Hybrid III* for frontal impact collisions
- *Side impact dummy* (SID) for side impact collisions
- *Biofidelic rear impact dummy* (BioRID) for rear impact collisions

As injuries vary greatly depending on the occupants' sizes and masses, crash dummies must represent the whole bandwidth of possible occupants. Hence, Part 572 of Title 49 of the *Code of Federal Regulations* (CFR) specifies a whole family of Hybrid III dummies (see Table 7.9 and Fig. 7.73).

---

[26] However, researchers like Colonel John Paul Stapp from USAF or Prof. Lawrence Patrick from Wayne State University performed selftests in the early days of crash testing.

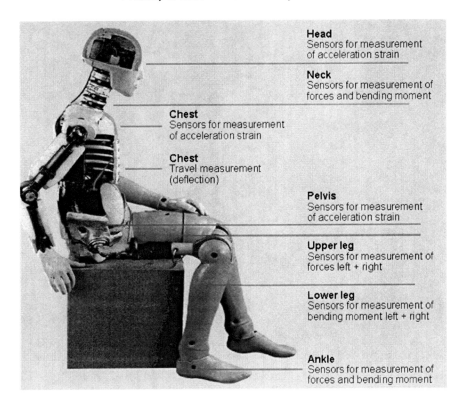

Fig. 7.72 Sensors in a Hybrid III crash test dummy (Source: FTSS)

Table 7.9 Hybrid III crash test dummy family specification (Source: CFR)

Type	Height	Weight	Seat position
50% male[1]	175 cm / 5 ft 9 in	78 kg / 170 lb	Driver (standard)
95% male	188 cm / 6 ft 2 in	100 kg / 223 lb	Driver (alternative)
5% female	152 cm / 5 ft	50 kg / 110 lb	Front pass. driver
Ten year old child		35 kg / 78 lb	Rear
Six year old child		21 kg / 47 lb	Rear
Three year old child		15 kg / 33 lb	Rear

[1] The value indicates the percentage of humans of the same gender who are smaller or equal in size and weight (for the US).

Fig. 7.73 Hybrid III family (Source: BMW)

The future successor of the Hybrid III is the THOR *advanced crash test dummy* which offers numerous functional benefits as compared with previously existing crash test dummy technology [40]:

- Assistance in optimization of "smart" occupant restraint systems, including setting of air bag deployment thresholds and design of integrated advanced belt/ bag restraint systems
- Improved assessment of belt/ bag interactions
- Improved head kinematics, as a result of new neck and spine designs
- Improved neck injury assessment, including out-of-position injury assessment
- More realistic spinal kinematics, as a result of added spine flexibility
- Facility for assessment of seated posture effects upon restraint performance
- Detection of air bag, belt, and wheel rim interaction with the abdomen
- Facility for injury assessment at the hip joint
- Detailed assessment of foot motions and ankle/foot/tibia injury potential
- Facility for localized facial injury assessment

Other dummy types – which are not part of any mandatory testing but are used in research to obtain insight on the respective injury processes – include e.g. the *child restraint air bag interaction* (CRABI) family representing infants at ages 6 months, 12 months and 18 months, or dummies representing pregnant women.

Test Process

Especially when performed to prove a vehicle's conformance with legal requirements, crash tests must precisely follow the specified procedures. These specifications can include detailed requirements for:

- The barrier (e.g. geometry, materials, construction, mounting etc.)
- Test setting (e.g. relative movement of vehicle and barrier, location of impact, speed at time of impact etc.)
- Test vehicle preparation, (e.g. loads and load distribution, setting/position of seats, armrest, handbrake, gear, filling of fluids etc.
- Body intrusion measurements before and after crash
- Dummy specification and preparation, e.g. dummy type, clothing/footwear, positioning etc.
- Preparation and usage of measuring devices
- Test photography (e.g. number, type and position of high speed cameras)
- Calculation of injury parameters and total rating

As part of the test preparation, the instrumentation of the crash test dummies has to be calibrated and checked for proper function. Next, target marks are attached to the side of the head to facilitate later analysis. After being placed in the test car, the data channels (up to 44 for the Hybrid III Family) are allocated and checked. Marking paint of different colors is applied to hands, knees and head to allow identification of the respective contact points in the car later on.

For the actual crash, the test is car pulled towards the barrier at the specified speed by means of a towrope which is disengaged just before the car hits the barrier. The overall duration of the crash phase only lasts about 100 to 150 ms. Figure 7.74 shows depicts the four main different types of crash tests: A BMW 3 Series in a frontal impact test against an offset deformable barrier, a BMW X5 in a side impact test, a MINI Cooper S in a side pole test, and a BMW 3 Series in a rollover test.

Documentation of the results is carried out on two levels: Film and data recording. Up to 20 high speed cameras film the crash scene from different perspectives, shooting around 1,000 frames per second each to create a detailed picture of the deformation and movement processes of the car and its occupants. At the same time, the instruments within the dummies record the injury data and store them in a temporary repository located in the dummy's chest or the vehicle's trunk from which they are downloaded after the crash. Between 30,000 and 35,000 data items are typically recorded per dummy. In addition, several triaxial accelerometers fixed to the vehicle body transmit acceleration data.

Fig. 7.74 Real crash tests: Frontal impact, side impact, side pole impact, rollover (Source: BMW)

7.8 Theft Deterrence

7.8.1 Legal and Customer Requirements

Even if the comprehensive section of an auto insurance policy might cover theft (see Sect. 7.2.1.1) and eventually make good the related financial damage: Having his or her car stolen is among the biggest possible nuisances for any automobile owner. Nevertheless, protection from theft, as a complete vehicle characteristic, only gets a minor share of attention, both from the customer and in development. While agility, comfort or passive safety are part of nearly every sales conversation, customers asking about the level of protection a car offers against theft are in the minority. In the same way, vehicle development teams usually include distinct members responsible for meeting the targets from the complete vehicle characteristics mentioned before, but rarely name a responsible person for theft protection.

For this book, the term *vehicle related theft* is used for all of the three different types of larceny that can occur to someone in consequence of owning a car, namely:

- Theft of the complete vehicle (grand theft auto)
- Theft of vehicle components
- Theft of personal items stored in the vehicle

For all three types of theft, the statistical risks of becoming a victim has four main parameters:

- The worth and attractiveness of the vehicle, its components or stored items
- The technical hindrances the car poses to a potential thief
- The risks for a potential offender of being traced back
- The region (e.g. district) and location (e.g. garage) where the car is parked or driven

While the last of these parameters is determined by social factors such as wealth of people or the availability of safe structures, the first three parameters are determined solely by the design of the car. As the motivation level for every thief can be quantified as the ratio of the subjective value of the expected gain to the potential punishment if the attempt fails, the two strategies for design for theft prevention are clear:

- Making the vehicle and its components as unattractive to thieves as possible
- Making the theft process as difficult and risky as possible

These two general strategies obviously impose conflicts with other targets, e.g.:

- Attractiveness to the customer vs. attractiveness to the thief
- Ease of authorized cabin access vs. inhibited unauthorized cabin access
- Quick access to the cabin for rescue of personnel in case of an accident vs. inhibited unauthorized cabin access
- Ease of operation vs. unauthorized operation
- Ease of replacement of lost keys vs. unauthorized copying of keys
- Ease of replacement of components vs. unauthorized removal of components
- Tamper-proof anti theft devices vs. risk of erroneous immobilization

How different measures for theft protection solve these conflicts will be discussed in Sect. 7.8.2.

7.8.1.1 Theft of Complete Vehicle

Due to anti-theft devices like electric engine immobilizers, auto theft is fortunately on the decline worldwide. In Germany e.g., where electric engine immobilizers are mandatory since 1998, the theft rate has decreased by 80% between 1993 and 2006 according to the *German Insurance Association* (GDV). Similar figures are valid for the United Kingdom or Australia, where immobilizers are likewise legally required. In the U.S., where these devices are available but not mandatory, theft rate for complete vehicles went down by 19% from 1995 to 2006 [41].

There are four main reasons cars are stolen:

- To be driven for fun (joyride) and abandoned thereafter
- To be used in a crime and abandoned thereafter
- To be driven by the thief him- or herself
- To be sold as a whole or chopped in components

According to statistics from Australia, the first two reasons account for 75% of all car thefts [42]. Keeping a stolen vehicle for yourself requires intensive effort to counterfeit documents and manipulate identification tags and therefore is the exception. For the same reason, it is difficult to resell stolen cars within the country where they have been stolen. Hence, cars which are stolen for the purpose of reselling them are typically illegally shipped to countries without reciprocity for registration. This is often done by criminal organizations originating in these countries who take orders for the cars they steal.

In order to take the appropriate technical countermeasures, the theft process must be analyzed. Here, a decisive factor is, whether the potential thief possesses the car key as the means to open and start the car or not. If not, there are two possible ways to steal a car:

- Gaining forced access, disabling the steering lock, hotwiring the engine
- Towing the vehicle away

However in the majority of cases today, the offender is in possession of the vehicle's key – which obviously enables him to drive off with it. The typical ways to get possession of a vehicle's key for the purpose of stealing it are:

- Removal of a key without the owner's consent through persons who have access to the keys
- Removal of a key the operator has carelessly left unattended
- Steeling or copying a key
- Taking a vehicle (with key) by force (carjacking)

A completely different way of illegally taking possession of a car is fraudulent vehicle theft, which means paying for a car by means of a fraudulent transfer of funds that the seller will ultimately not receive.

7.8.1.2 Theft of Components and Personal Items

Thanks to the widespread use of advanced anti-theft technology, complete vehicles are no longer easy prey. Given this background, more importance now has to be placed on theft of components and of personal items – which both have to be looked at separately in order to select the right measures to fight them:

- Costly vehicle components – in the majority if cases – are aimed for by professional criminals who work by order from a third party. They seek exactly for the part they need, possess sophisticated technical knowledge and use professional tools. On their hit lists are air bags, xenon head lights, navigation systems, stereo systems, and – due to the content of precious metals such as platinum and palladium – catalytic converters. In their "working style", avoiding public attention is more important than time needed.
- Personal valuables however are typically sought-after by rather inexperienced individuals who urgently need money – in many cases to buy drugs – and try to get it by selling the stolen items as quickly as possible for cash. This type of criminal looks for anything that seems to be sellable quickly (such as cell phones, cameras, stand-alone navigation systems etc.), and try to access it fast – e.g. by breaking a window or ripping a soft top.

7.8.1.3 Legal Requirements on Theft Deterrence

In basically all markets legal requirements exist that should provide a certain level of security from vehicle related theft. As insurance companies bear the biggest share of the related financial burden, their representative bodies such as the *Gesamtverband der Deutschen Versicherungswirtschaft* (GDV) in Germany, the *Motor Insurance Repair Research Centre* (MIRRC) – commonly called

Thatcham – in the UK, or the *American Insurance Association* (AIA) in the U.S. are the driving force behind these legal requirements.

Worldwide, there are mainly two different standards specifying the legal requirements for theft deterrence: FMVSS/CMVSS 114 and ECE R 116.

FMVSS/CMVSS 114

For type approval in the U.S. or Canada, compliance with FMVSS/CMVSS regulations is required (see Sect. 7.1.3). The requirements in terms of theft prevention are covered by FMVSS/CMVSS 114, and are in general [43,44]:

- A key-locking system that, whenever the key is removed, prevents the normal activation of the vehicle's engine and either steering or forward self-mobility of the vehicle or both.
- A warning to the driver activated whenever the key to the key-locking system has been left in the locking system and the driver's door is opened.

Additional requirements that apply to vehicles with automatic transmission ensure safe operation of the vehicle:

- The key may only be removed if the transmission is locked in "park".
- When locked in "park", vehicles must not move more than 150 mm on a 10% grade
- The vehicle's steering or forward self-mobility may only be deactivated when it is in "park".

ECE R 116

ECE R 116 both lists up and specifies the theft deterring devices which are mandatory according to UNECE Type Approval (compare Sect. 7.1.3), meaning they are binding for basically all countries other than the U.S. and Canada. Stated requirements include [45]:

- Devices that prevent unauthorized starting of the engine (e.g. ignition lock), and prevent unauthorized steering, driving or moving forward under its own power (e.g. steering lock, transmission lock, interlock). The required technical specifications for these devices (such as minimum numbers of key combinations, lock technology, resistance to force, etc.) are included in part I of ECE R 116.
- An electronic engine immobilizer that disables at least one vehicle circuit needed for vehicle operation under its own power (e.g. starter motor, ignition, fuel supply, pneumatically released spring brakes, etc.). If the vehicle is equipped with a diesel engine or the immobilizer has been fitted after purchase, at least two separate vehicle circuits must be disabled. The required technical specifications (such as coding technology used for setting and unsetting,

operating reliability and safety, replacement of components, etc.) are included in part III of ECE R 116.

Although installation of alarm systems is not required by ECE R 116, it specifies the technical requirements for these systems and vehicles equipped with them (such as the extent of intrusion detection, cabin surveillance technology, duration and intensity of audible and visible alarm signals, safety against false alarm, etc.). These specifications are included in part II of ECE-R 116.

7.8.1.4 New Vehicle Security Rating

While FMVSS/CMVSS 114 and ECE-R 116 specify certain required functions, the *New Vehicle Security Rating* (NVSR) procedure issued by the *Motor Insurance Research Repair Centre* (MIRRC) – usually referred to as *Thatcham* – allows an objective quantitative evaluation of a vehicle's level of theft protection – similar to the passive safety assessment according to NCAP (compare Sect. 7.7.1.4). The assessment evaluates the security level in three categories [46]:

- Resistance to forced entry
- Resistance to driving away
- Component identification and replacement

The maximum of 1000 points that can be allocated by a tested vehicle, are distributed over different categories as follows:

- Access to passenger compartment (max. 175 points):
  Assessed by means of a standardized attack test. The tested vehicle must resist forced entry to cabin, hood and trunk for at least 2 min. Wheels must be secured.
- Override ignition and/or steering column lock (max. 90 points):
  Assessed by means of a standardized attack test. The tested vehicle must resist attempted manipulation of the locks for at least 2 min (in addition to the time needed to access the cabin).
- Vehicle and component parts identification (max. 115 points):
  Assessed by evaluation of which parts are identified with an unambiguous code and how safe these identifiers are from being manipulated.
- Availability of compliant electronic immobilizer (max. 400 points)
- Availability of compliant alarm system (max. 200 points)
- Security of key and component replacement procedures (max. 20 points)

The total number of accumulated points is then – in analogy to the NCAP rating – converted into star ratings. A maximum of five stars is awarded each for the effectiveness of measures against "theft of" and "theft from" a car.

7.8.2 Component and System Design

Based on the different theft types discussed above, Fig. 7.75 gives an overview of theft processes and is used as a basis for identifying countermeasures.

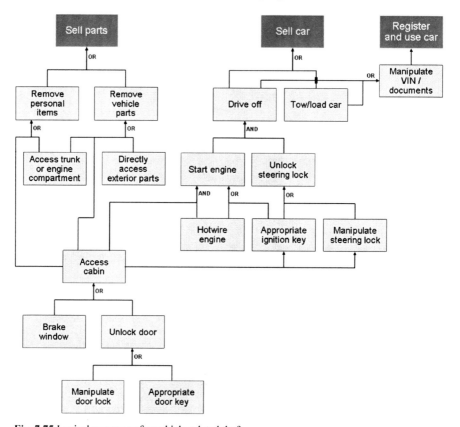

Fig. 7.75 Logical sequences for vehicle related theft

Six different theft prevention strategies can be derived from these processes:

- Inhibiting unauthorized vehicle removal
- Inhibiting unauthorized access to cabin, trunk, or engine compartment
- Inhibiting unauthorized operation of the vehicle
- Diminishing value or usability of stolen items
- Lowering the general theft risk

7.8.2.1 Inhibiting Unauthorized Removal

Pulling a vehicle or loading it on a flatbed truck requires the respective equipment, but can be done relatively quickly and – depending on the circumstances – relatively inconspicuously. Sensors of integrated theft alarm systems however detect irregular movements of the vehicle that set off an alarm or wirelessly inform the owner.

The most effective way of preventing auto theft however would be to physically inhibit the vehicle's removal. But while e.g. bicycles or motorcycles are often locked to immovable items such as utility poles or steel grids, this approach has been practically never followed with a car – apart from parking it in a protected environment (which is independent from the vehicle's design). In addition to technical difficulties, possible solutions are further restricted by legal regulations that require means for physically inhibiting vehicle dislocation to allow for necessary removal of a vehicle e.g. by parking enforcement or to create access for emergency vehicles.

7.8.2.2 Inhibiting Unauthorized Access

If the target is not a component that can be accessed from outside (e.g. wheels), the first thing any offender has to do is to get access to the vehicle's interior – or likewise the engine compartment. As discussed above, there are two significantly different motivations and approaches to unauthorizedly access a vehicle: If the target is to steal and use or resell the car, damaging it would diminish its value and hence be minimized. If however the target is to dismount components or to take possession of personal items, damage to the car is of no importance to the thief. All that counts is to get access very quickly. For the owner then, the damage to the car is often much higher than the loss incurred by the stolen objects. Table 7.10 lists established features that prevent unauthorized cabin access.

Table 7.10 Systems inhibiting unauthorized cabin access

System	Possible security features
Locking system (doors, hood, trunk lid)	Cylinder and locking mechanism are protected against mechanical manipulation (e.g. drilling through the body shell from outside)
	Remote control signal can not be cloned
	Locking system is protected against electronic manipulation (e.g. after access to the respective bus system has been gained)
	If vehicle is locked, doors, hood, trunk lid, can not be opened from the inside (double lock)
Windows and sunroofs	Mounting of shatter proof glass
	Regulator system is protected against electronic manipulation
Soft top	Cutting and ripping proof structure and fabric
	Opening/closing system is protected against electronic manipulation
Alarm system	Detects mechanical tampering with the locking system
	Detects attempted bus intrusion
	Detects motion inside the cabin

7.8.2.3 Inhibiting Unauthorized Operation

As the mandatory steering column locks were usually easy to break for an adept thief, for a long time additional steering wheel locks were the preferred means to prohibit unauthorized operation. Especially in high theft-risk areas and for high theft-risk cars, car owners purchased massive steel locks that mechanically prohibit turning of the steering wheel. The major weakness of this practice however is that the lock needs to be applied and secured manually. But just as a seat belt that is not worn can not save lives, a steering wheel lock that is not engaged does not prevent car theft.

Given this background, one of the reasons for the success of electronic engine immobilizers discussed above is that they are actually engaged and disengaged without driver awareness. Immobilizers that meet legal requirements today interrupt the power to three or more electronic circuits, usually the starter motor, ignition and fuel pump circuits. Only when the specially coded key or token is put in the ignition switch can these vital circuits work. Thus, the vehicle can only be operated with the original coded key or other coded element.

7.8.2.4 Diminishing Value or Usability of Stolen Items

In addition to the measures discussed above which physically obstruct the theft process, rendering the vehicle and/or specific components useless for the thief is another approach of theft deterrence. The vehicle and its components can be outfitted with features that result in illegal use or resale requiring too much effort or posing a too high risk to the offender.

An unambiguous mark by which authorities can identify a vehicle's legal owner is the *vehicle identification number* (VIN). A common approach to protect the VIN from being manipulated is to apply it to a part of the body behind the lower edge of the front screen, requiring disassembly and potential destruction of the front screen to access the VIN. In addition, parts that require a high effort to replace (e.g. windows) can be permanently marked with the VIN (see Fig. 7.76). In the same way, invisible identification tags that only can be read e.g. in ultraviolet light can be applied to the car. Manipulation of the VIN – as required for inconspicuous registration – is then extremely difficult and thus makes the car relatively worthless for resale. A prerequisite for the desired deterring effect nevertheless is that the potential offender is aware of this security feature by clearly visible signs.

Fig. 7.76 VIN etched on a side window as theft deterrence
(Source: Securmark)

Likewise, valuable vehicle components such as rims can be furnished with the VIN as a means of theft deterrence through retraceability.

An effective means to quickly locate stolen vehicles and get them back to their owners are *vehicle tracking systems*, also referred to as *after theft systems for vehicle recovery* (ATSVR). These systems identify a stolen vehicle and its location and continuously send this information to a privately operated security center or to the police.

An enhancement to ATSVR systems are *after theft vehicle immobilization systems*, which help recover stolen vehicles in two different circumstances [46]:

- Slowing down and then stopping a moving vehicle in a controlled manner by authorized personnel
- Preventing the engine of a stolen vehicle from being restarted once it has been stopped

The major benefit of these systems however is not so much the higher likelihood of getting a stolen vehicle back, but the lower likelihood of getting the car stolen in first place – due to the deterrent effect on potential thieves.

7.8.2.5 Lowering the General Theft Risk

With ever improving anti-theft systems the driver of the car is also a critical element of theft prevention. If they would carefully follow the security advice given by police, government, or insurance companies, vehicle related theft could be further reduced. But – be it out of negligence or ignorance – many car owners forget to alarm their vehicle, leave windows open, park in unknown dark areas, or engage in other unsafe practices.

Many features available today at least partially compensate for these human weaknesses. Acoustic warnings e.g. can be given if the car is locked and windows are not properly closed, if an open convertible is parked and any unsecured item is sensed in the cabin, or if the driver after parking the vehicle opens the door with the key still in the ignition lock.

A major risk factor for auto theft is – as discussed above – the area where the car is driven or parked. Here, navigation systems can provide useful information. A concept from Honda e.g. is to provide the driver with information about areas with high theft or vandalism rates and in doing so giving him or her the opportunity to avoid risky areas.

7.8.3 System Integration and Validation

As theft usually occurs unobservedly, information about how theft attacks are really performed is generally unavailable – especially for successful ones. Hence, direct evaluation of how well the design of a vehicle and its respective devices prevent theft of or theft from the vehicle is not possible.

A realistic assumption of theft attempts however are the attack tests specified by Thatcham. The related standardized procedure allows quantitative evaluation and hence comparison of different cars in terms of theft deterrence, the aforementioned NVSR [47]. Figure 7.77 shows an attack test carried through at Thatcham Laboratories.

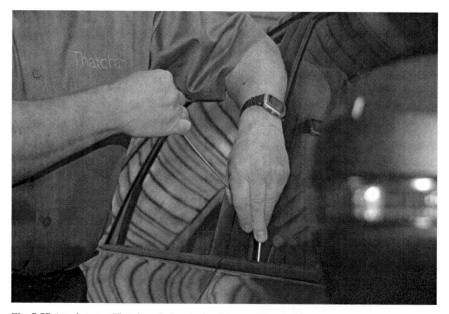

Fig. 7.77 Attack test at Thatcham Laboratories (Source: Thatcham)

7.9 Reliability

7.9.1 Legal and Customer Requirements

7.9.1.1 Reliability vs. Quality

Product quality as the level of fulfillment of customer needs has already been discussed in Sect. 6.4, along with the respective processes. An additional important requirement to a complete vehicle however is reliability, the ongoing fulfillment of these requirements over the whole life span of the vehicle.

Although being among the first things customers considering purchasing a used vehicle think about, reliability on a complete vehicle level is rarely even mentioned in books on vehicle development – a phenomenon for which there are probably two reasons. Firstly, reliability is obviously not as important for new car buyers, who definitely are the focus group of vehicle development. Secondly, a structural change has occurred over the last years in the technical patterns that lead to component, system and vehicle failure: While formerly complete vehicle reliability was mostly determined by the reliability of its components (e.g. body, engine, transmission), today the majority of reliability problems the owner of a vehicle experiences are due to malfunction of software controlled systems. The ADAC's analysis of vehicle breakdowns in Germany shown in Fig. 7.78 supports this clearly.

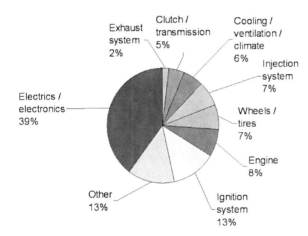

Fig. 7.78 Sub-systems causing vehicle breakdowns
(Data source: ADAC [48])

7.9.1.2 Measuring Reliability

In general, reliability is defined as the ability of an item to – or likewise the probability that an item can – perform a required function under given conditions for a given time interval [49,50]. The metrics that are typically used to specify reliability requirements of technical systems such as vehicles are the *mean time between failures* (MTBF) or its reciprocal value, the *constant failure rate* (CFR). For passenger cars, the MTBF is usually specified in miles (or kilometers) driven.

For so-called single-shot-systems such as airbags etc. that remain dormant during normal operation and only need to operate once, declaration of an MTBF or failure rate does not make sense. In this case, reliability is measured by the systems' probability of success.

How meaningful the declaration of a reliability figure such as MTBF is, depends however on how realistically the "given conditions" in the definition above are considered. The big challenge in reliability engineering is to anticipate the way a vehicle will be used and take this assumption as the basis of the design and dimensioning process, e.g.:

- Mechanical and thermal load patterns: Is the vehicle usually driven gently or sportily? Is it driven on rough or even surfaces? Is it used for short range or long range drives? Is it usually fully loaded or driven with only the driver's weight as additional vehicle load?
- Environmental patterns: Is the vehicle is rather driven in a dry, hot, and dusty inner-city environment, on wet, cold, and salty roads, or in a moderate climate?
- Operational patterns: Which functions do the driver (or the passengers) use? Which functions are used at the same time? How do driver and passengers operate these functions?

Compared to the generic definition of reliability given above, the potential customer of a passenger car usually has a much more concrete idea of what reliability of her or his vehicle means, e.g.:

- How often will the car have a breakdown and need to get towed?
- How often will I have to drive the car to a dealership to have it repaired?
- How often will I have to drive the car to a dealership for provisional maintenance?

7.9.1.3 Reliability of Embedded Software

A component that follows completely different rules in terms of reliability is software. In complex electromechanical systems like passenger cars, embedded software has become the determining factor for system reliability. In most cases, it is neither the mechanical parts like body or engine, nor the actuators, sensors, harnesses, nor the ECU hardware that cause the overall system to fail; it is software that instead of fulfilling the desired function, creates an unwanted or unexpected system state leading to malfunction of subsystems or the complete vehicle (see Sect. 5.1.1).

According to the *American National Standards Institute* (ANSI), software reliability is defined as the probability of failure-free software operation for a specified period of time in a specified environment [50]. The aforementioned reliability metrics – such as MTBF or CFR -historically stem from the observation of hardware systems and do not make much sense when applied to software. A suitable metric for software reliability is e.g. *fault density* (FD), usually measured in number of faults per thousand lines of code. Table 7.11 shows typical target values:

Table 7.11 Accepted failure rates for different types of software (Source: Carnegie Mellon University [51])

Software type	Failure rate [per 1000 lines of code]
Normal software	25
Critical software	2 to 3
Medical software	0.2
Space shuttle software	< 0.1

7.9.1.4 Severity of Failures

In addition to availability, the severity of failures is another aspect of reliability. It summarizes the possible consequences of a particular failure. An excerpt of the problem evaluation table shown in Fig. 6.4, Table 7.12 introduces a useful metric for measuring severity:

Table 7.12 Levels of failure severity (Source: BMW)

Severity level	Consequences
5	Can be repaired at next opportunity
4	Requires urgent visit to workshop
3	Breakdown, vehicle needs to be towed
2	Loss of mandatory properties (vehicle becomes illegal)
1	Vehicle poses life threatening risks to driver or environment

7.9.1.5 Reliability Related Costs

In addition to the risks and inconveniences lack of reliability poses to the owner of a vehicle, it also represents a major cost factor. How reliability costs contribute to total costs has already been discussed in Sect. 7.2.1.2. This is especially true for commercial customers – like vehicle fleet managers or rental car agencies – where the vehicles' availability directly affects the economic performance of the business. The vehicles' availability is of utmost importance, MTBF or CFR targets are part of their business plans and contracts.

But lack of reliability does not only contribute to the owner's cost. Quite apart from the indirect financial damage a manufacturer suffers through loss of reputation and unsatisfied customers, warranty costs in the automotive industry total over US$ 10 billion per year, representing more than 2% of its product revenues. For an industry with a typical average profit margin of US$ 175 per vehicle, the resulting warranty costs of US$ 700 per vehicle represent long term liabilities and an extreme financial pressure. To make things worse for the manufacturers, markets today demand ever longer and more comprehensive warranties [52].

7.9.2 Component and System Design

Design for reliability means that potential sources for failure must be eliminated from the very beginning and follows – like all other complete vehicle characteristics – the V-model introduced in Sect. 1.3.3: Reliability targets are agreed for the complete vehicle, then designed in on a component level, and then validated on component, sub-system and system level. This requires both knowledge about general failure processes of different types of components, and knowledge about the reliability related structures of the complete system that determine how component failures are propagated into system failures.

7.9.2.1 Component Level Reliability

In the automotive world, the parameters that determine a vehicle's reliability have changed over the last years: While traditionally corrosion of body and chassis parts, and wearing out of drivetrain components were the leading defects, reliability today is mostly a matter of software controlled E/E systems (see Sect. 5.1). The effects for the customer might be similar, but the underlying causal chains are completely different and must be understood on a detailed level.

Failure of Mechanical and Electrical Components

A load related failure denotes malfunction of a part through wear, aging or over-load, e.g. the fatigue fracture of a compressor drive shaft leading to failure of the air conditioning system, or the failure of a semiconductor in an engine ECU leading to a breakdown of the engine etc. Based on an assumed load spectrum and multiplied by the applicable factor of safety, the required operational strength for a component is calculated and documented in the component's formal specification. The required operational strength is the basis for the dimensioning of the component as part of the design process, and conflicts with other targets such as material costs, available space and mass/moment of inertia.

A good example for this approach is the dimensioning of chassis components. As shown in Fig. 7.43, the load spectrum of a vehicle in normal everyday use differs greatly from the one a vehicle is exposed to when driven on a race track. But as chassis components are critical to safety, it is the race track load spectrum that is taken as the basis for dimensioning the components to lifetime durability, even though the normal requirements concerning operational strength are much lower.

With this process in mind, there are two basic causes for failure of hardware components:

- The real load spectrum exceeded the loads assumed by design:

  - The assumed load spectrum was unrealistic (e.g. the assumption concerning how often a convertible roof will be opened and closed over lifetime)
  - The real loads exceeded the loads design could assume for normal operation (e.g. driving over a curbstone at more than walking speed)

- The component was weaker than expected:

  - The dimensioning process was incorrect (e.g. application of inaccurate simulation algorithms to calculate a part's dimension)
  - The component has not been manufactured or assembled to specification (e.g. the number of pores in a cast tension strut exceeds the tolerated level)
  - The component has been pre-damaged during production or transport (e.g. oscillating stress on wheel bearings in standstill during a vehicle's sea transport)

As part of their design process, the fatigue behavior of singular mechanical components can be simulated using the *finite element method* (FEM). Taking a tension strut as an example, the left picture in Fig. 7.79 shows the result of the numerical simulation predicting an early crack initiation after 450 to 600 load changes in the red colored area right under the upper eye. The right picture shows the initiation crack the real tension strut suffered in testing after 292 load changes.

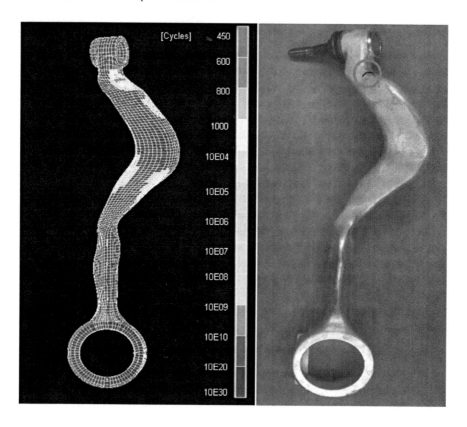

Fig. 7.79 Fatigue crack: Simulation and real test result (Source: BMW)

If measured over time, the failure rate of conventional mechanical or electrical systems shows a distinct behavior known as the bathtub curve (see Fig. 7.80). It can be divided in three main phases:

- Failures in brand new systems (phase I) are typically caused by lack of product maturity or production faults. As the respective problems are solved both product and process are getting more and more mature, and therefore the early phase can be considered the last phase of product development – with the product already in the customer's hands. The more the vehicle has been designed for quality and the more it has been tested before delivery, the lower the starting point of the bathtub curve during this phase. A mature system that has been built and improved for years has practically no early phase failures.
- For the major part of the system's lifetime (phase II), random failures occur resulting in a constant failure rate that is as low as possible. Phase II should last at least as long as the first customer usually keeps the product.
- Later in the lifetime of the system (phase III), wear and fatigue failures lead to an increasing failure rate. This phase should not start before the expected lifespan of the product is over. With the increasing failure rate, repair costs

increase. When repair costs exceed the actual value of the product, it is de-commissioned.

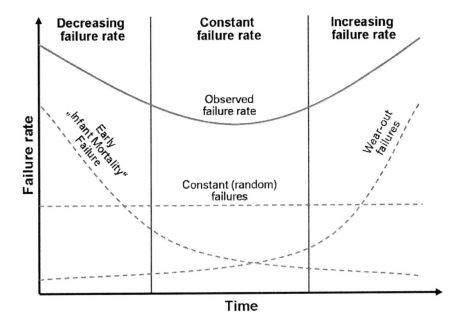

Fig. 7.80 Bathtub curve: Failure rate over vehicle lifespan

Electronic Components

Compared to mechanical components, electronic components (e.g. processors, memory chips, or general semiconductors) show a slightly different failure pattern. The three main factors leading to malfunction are:

- Ambient temperature: In most cases, the actual root-cause for microprocessor failures is failure of the cooling system.
- Workload: The dependability of microprocessors depends highly on the actual workload.
- Power stability: The majority of memory failures are soft-failures, caused by variation of the supply voltage.

Embedded Software

Software fails in a completely different way than hardware does. Since hardware failure is caused by physical or chemical processes, the overall system then can be brought back into the original working state to perform its intended function by repairing or replacing the defective component. Software however does not age or wear out. Software failure – that is the execution of an unintended operation – is the result of an unanticipated combination of operational states and inputs, and is always design related. In contrast to most hardware failures, software failures always happen without any prior warning.

Typical reasons for software failure are [53]:

- False or misinterpreted specification of the software's requirements
- Carelessness or incompetence in writing code
- Inadequate testing of the software
- Incorrect or unexpected usage of the software

As – even in relatively small programs – the number of possible input/state combinations can be extremely high and extremely time-consuming to test, following a systematic software development process is an indispensable element of software reliability engineering.

Over time, the failure rate of software follows a different pattern than hardware does. The revised bathtub curve for software reliability shown in Fig. 7.81 demonstrates the adverse effects of regular functional updates during the main usage phase, and the flat end that comes from software not being subject to aging or wear-out [51].

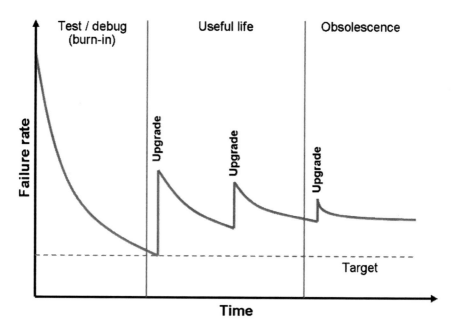

Fig. 7.81 Revised bathtub curve for software (Source: Carnegie Mellon University [51])

The first step in creating reliable software is to specify the program's desired functions in a structured way (see Sect. 5.2.5.2). Classical stochastic software reliability models that structure the software failure process are able to estimate the current reliability and to predict future failure behavior. A metric that allows a quantitative reliability evaluation based on the analysis of the vocabulary used in the requirement statement counts lines of text, imperatives, continuances, directives, weak phrases, incomplete statements, and options [54].

Based on the requirements specification, software design then determines how the program should fulfill the specified functions. A well structured high-level design process specifies architecture independently from coding and is an indispensable prerequisite for the creation of reliable software. At this stage, size and complexity of the architecture are the two main metrics for reliability.

The actual programming is then done in an advanced programming language, for automotive application usually in C. Code analysis methods produce metrics that quantify the reliability of the program. An important programming requirement in terms of reliability is to ensure that further development of the program as necessary during the useful life phase (see Fig. 7.81) is possible without creating additional errors.

Electro-magnetic Immunity

A separate family of failure types is malfunctions caused by unintended or ambient electro-magnetic fields. Of the two aspects included under the term *electro-magnetic compatibility* (EMC), immunity against electro-magnetic fields is the one relevant for a vehicle's reliability (the other aspect, electro-magnetic emissions, is discussed in Sect. 7.10.5). Examples of EMC related failures are [55]:

- Windshield wipers can be "heard" in the radio
- Certain FM radio stations can't be received while the headlights are on
- Engine misfires when driving under a high-voltage power line
- Car is disabled and engine controller damaged by operating an amateur radio transmitter installed in the trunk
- Electrostatic discharge occurs while inserting the ignition key, damaging ignition circuitry
- Nearby lightning strike causes cruise control to engage and accelerate

An analysis of EMC related problems brings up three main failure modes:

- Disturbance of audio and video signals
- Destruction of electronic elements through induced high voltage impulses
- Unwanted and incalculable operations through voltage impulses induced in cables or electronic elements that are misinterpreted by an ECU as an input signal

A modern passenger car has about 70 highly interconnected ECUs, up to 500 sensors and actuators, about 1.9 miles of wires, firing pulses up to 40 kV, and about

15 internal antennas that detect voltages of a few µV between 125 kHz and 2.8 GHz. It is thus very sensitive to electro-magnetic perturbation. Generally speaking, the higher the field intensity, the bigger is the risk potential to the vehicle.

Sources for unwanted electro-magnetic fields are both part of the vehicle and the vehicle's environment. Internal sources that hence can be influenced by the vehicle's design are:

- Pulse modulated high power currents as used e.g. in engine controllers
- Electrostatic discharge of occupants
- User installed radio transmitters
- Power switches

From a vehicle designer's point of view, external electro-magnetic fields must generally be accepted as they are; the design can only focus on immunity to these perturbations. The main external sources for electro-magnetic fields are:

- Electro-magnetic emissions from other vehicles (see internal sources)
- Radio and TV networks
- Power lines
- Lightning

Design for EMC is a bottom-up process. First, the solitary integrated circuits must be optimized. Here, micro-controller chips and bus drivers (e.g. CAN transceivers) are of primary interest. Then, complete ECUs are investigated. Figure 7.82 shows an ECU together with the relevant harness section undergoing a strip line immunity test according to ISO 11452-5 [56].

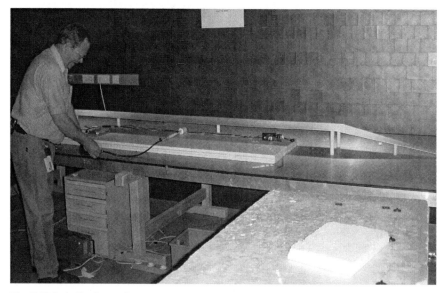

Fig. 7.82 Strip line immunity test according to ISO 11452-5 (Source: BMW)

Strip-line immunity tests are e.g. relevant in testing audio equipment for immunity to interference e.g. from cell phones' GSM type modulation. In terms of disturbance caused by an ECU, the FM frequency range is of primary interest. Models of the printed circuit board and the micro controller allow early simulation of the controller's emission behavior with sufficient accuracy. Figure 7.83 compares the simulation results with that measured at the real ECU.

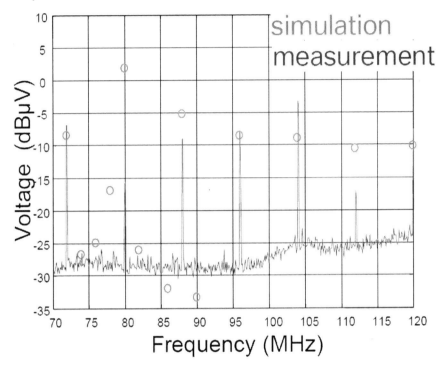

Fig. 7.83 Comparison of simulation and measurement of ECU emissions (Source: BMW)

General measures to improve electro-magnetic immunity are usage of shielded housings and lines, grounding, and decoupling of entries through filters (resistors, capacitors, chokes). According to [57], the four most important EMC guidelines for printed circuit board design are:

- Minimize the loop areas associated with high-frequency power and signal currents
- Don't split, gap or cut the signal return plane
- Don't locate high-speed circuitry between connectors
- Control signal transition times

7.9.2.2 System Level Reliability

The reliability of the complete vehicle is determined by the reliability of its subsystems, which in turn is determined by the reliability of its components. The causal relationships however are not unique; system complexity – and thus potential unreliability – stems from components and systems mutually affecting each other's behavior. A failure of the thermal engine management e.g. can cause an unwanted increase of the engine compartment temperature, leading to failure of the engine controller.

On a system level, design for reliability starts with the creation of reliability models that represent the logical dependencies between the failure modes of subsystems and components. These models are the basis for the application of reliability methods.

The most common inductive method for bottom-up risk assessment in the automotive development process is the *failure mode and effects analysis* (FMEA). For one sub-system or component, an FMEA, lists all the failures modes that could possibly happen, and then assesses the consequences (effect) of each failure. In an extended version of the FMEA, the so-called *failure mode and effects criticality analysis* (FMECA), the listed failure modes are rated in terms of their criticality to the operability of the system or component.

To ensure quality and consistency of FMEA / FMECA risk assessments, a considerable number of standards have been developed through different industry sectors, especially in aviation, aerospace and nuclear power. For the automotive industry, established standards are e.g. SAE J1739 , IEC 60812, BS 5760 and VDA 96. Figure 7.84 shows a system FMEA form compliant with SAE J1739, created by the Relex Reliability Studio's FMEA/FMECA module (see below).

In addition to forms and general procedures, FMEA / FMECA standards provide instructions and recommendations for the quantitative evaluation of risk factors. Table 7.13 provides the ranking numbers for severity, occurrence probability and detection probability.

Obviously, the value of an FMEA depends largely on whether the pursuing team takes all possible failure modes into consideration, even the primarily unthinkable ones. As part of the aforementioned software tools, failure mode libraries support the creative process of predicting possible failure modes for a given component.

POTENTIAL
FAILURE MODE AND EFFECTS ANALYSIS
(DESIGN FMEA)

Relex

Level of Analysis: System

Model Year(s)/Vehicles(s): 2008 / Automobile

Core Team: John Doe, Jane Smith

Design Responsibility: Thomas Bickerton

Key Date: 11/07/2009

FMEA Number: 145-36

Page   1   of   1

Prepared By: Frank Sherman

FMEA Date (Orig.) 10/07/2009   (Rev.)

Item / Function	Potential Failure Mode	Potential Effect(s) of Failure	S e v	C l a s s	Potential Cause(s)/ Mechanisms of Failure	O c c u r	Current Design Controls	D e t e c	R. P. N.	Recommended Actions	Responsibility & Target Completion Date	Action Results				
												Actions Taken	S e v	O c c	D e t	R. P. N.
Disabled vehicle or loss of performance	Engine stall during parking maneuver	Disabled vehicle in line of traffic	10	YC	a) Microcomputer malfunction b) Power steering pressure switch failure	2	Display	4	80	Implement limited operation strategy	Lamar Carlson					
	Engine stall	Disabled vehicle	10	YC	Microcomputer malfunction	2	Display	4	80	Implement limited operation strategy	Lamar Carlson					
	Engine stumbles	Loss of performance	8	YS	HEGO sensor damaged	3	Revert to underspeed mode	3	72	None	Inga Stevenson					
	Low idle with air conditioning on	Rough engine operation	6	YS	Neutral drive switch failure	3	Display	3	54		Inga Stevenson					
	Engine misfire	Loss of performance	5	YS	Exhaust valve	2	Display	3	30	Implement limited operation strategy	Sonia Mathers					
	Engine rattle	Annoyance to user	2		Fuel quality or supply	4	None	2	16	Fuel system inspection	Lamar Carlson					

Fig. 7.84 System FMEA form (Source: Relex)

Table 7.13 Criteria for product FMEA rankings (Source: VDA [58])

	Ranking for the Severity S		Ranking for the occurrence probability O	Assigned failure portion in ppm		Ranking for the detection probability D	Certainty of the detection procedure
10 9	Very high Safety risk, non-compliance with legal regulations, break-down	10 9	Very high Very frequent occurrence of the failure cause, unusable, unsuitable design concept	500,000 100,000	10 9	Very slight Detection of a failure that has occurred is unlikely. Reliability of the design was not or cannot be proven. Detection procedures are uncertain.	90 %
8 7	High Vehicle operability is severely restricted, immediately stay in a repair shop is urgently required. Degraded function in important operating and comfort systems.	8 7	High Failure cause occurs repeatedly, proble-matic, immature design.	50,000 10,000	8 7	Slight Detection of a failure that has occurred is less probable. Reliability of the design can probably not be proven. Detection procedures are uncertain.	98 %
6 5 4	Moderate Vehicle operability is restricted, immediate visit to a repair shop not absolutely necessary. Degraded function in important operating and comfort systems.	6 5 4	Moderate Occasional occurrence of the failure cause, suitable design with suitably mature design.	5,000 1,000 500	6 5 4	Moderate Detection of a failure that has occurred is probable. Reliability of the design could possibly be proven. Detection procedures are relatively certain.	99.7 %
3 2	Slight Slight vehicle function impairment, repair at next scheduled visit to a repair shop. Degraded function in operating and comfort systems.	3 2	Slight Occurrence of the failure cause is slight, proven design.	100 50	3 2	High Detection of a failure that has occurred is very probable, confirmed by several mutually indepen-dent detection procedures.	99.9 %
1	Very slight Very slight function impairment, only detectable by specialists.	1	Very slight Occurrence of the failure cause is improbable	1	1	Very high A failure cause that has occurred is certain to be detected.	99.9 %

While an FMEA or FMECA analyzes the system bottom-up, a *fault tree analysis* (FTA) breaks down an undesired state of the system into boolean combinations of lower level events. By applying probability data derived from estimates or historical data to these lower level events, system level reliability predictions can be derived. While certainly not absolutely accurate, these predictions allow comparison of different design alternatives in terms of their reliability.

Development of the FTA methodology has mainly happened in the aerospace sector, driven by the need to clarify the circumstances that have led to catastrophic system failures like the explosion of the space shuttle Challenger in 1986. One of the most comprehensive descriptions of the FTA is the *Fault Tree Handbook* issued by NASA [59]. In general, the method is specified in IEC 61025 [60]. Figure 7.85 shows a sample fault tree for the unwanted event "accidental engine firing".

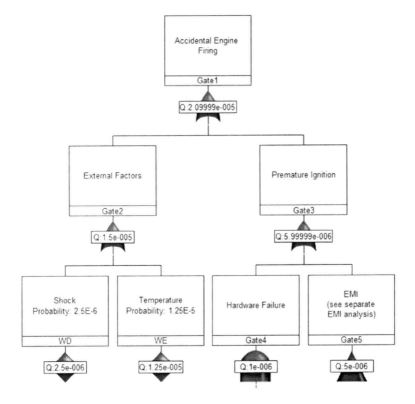

Fig. 7.85 Sample fault tree analysis (Source: Relex)

Creating FMEAs, FMECAs, and FTAs for complex systems is a complex task that requires proper project management and appropriate IT support. Established software packages for this purpose are e.g. APIS IQ Tools, Relex Reliability Studio, isograph Reliability Workbench, or Reliasoft Xfmea.

General design approaches that improve the reliability of a vehicle or its subsystems are e.g.:

- Redundancy of critical subsystems (e.g. brakes, lights)
- Decoupling of subsystems (e.g. no spreading of functions over several ECUs)
- Poka-yoke (fool-proofing of product or process operations)

The latter however plays its most important role not in the design of vehicle operation systems, but in avoiding manufacturing faults that could lead to vehicle failures.

As mentioned previously, production faults represent the other half of causes for vehicle failures. Deviations such as cavities in cast parts, bolts that are not tightened to the required torque, or contact surfaces of adhesive connections that have not been cleaned to specification, lead to reduced stability and can cause failure of the respective components. Anticipating and avoiding potential production problems over product development is within the focus of production integration, a secondary complete vehicle characteristic that is discussed in Sect. 8.1.

7.9.3 System Integration and Validation

7.9.3.1 Accelerated Testing

The central dilemma in complete vehicle reliability testing is that – if the results are to be incorporated in the design process – the behavior the series vehicle will show over its whole lifetime must be anticipated during development, at a stage where even the series production design state has not been reached. For this reason, accelerated testing methods are used during development.

Among the potentially damaging impacts a vehicle's components are permanently exposed to are environmental influences. Standardized environmental tests for components and complete vehicles such as the salt mist test according to IEC 60068-2-11, the flowing mixed gas corrosion test according to IEC 60068-2-60, change of temperature test according to IEC 60068-2-14, or the sunlight simulation test according to DIN 75220, allow vehicle development teams to bring about the long-term effects of environmental impacts in an accelerated way. As an example, Fig. 7.86 shows part of an instrument panel before (top) and after (bottom) a sunlight simulation test, with obvious deformation and discoloration caused by the intensive UV radiation.

An example for a comprehensive accelerated test program for complete vehicle corrosion is BMW's *dynamic corrosion test*. This intensive test procedure combines mechanical, thermal and chemical treatment of the car and is performed in three steps:

- Vehicle preparation: Anticorrosive coatings are locally scratched to the blank metal.
- Preconditioning: To create a specified level of pre-damage, the vehicle is stressed under extreme thermal, mechanical and chemical conditions, such as body torsion in cold climate, impingement of the vehicle with loose gravel, saltwater and sand, and driving over sinus waves, pot holes, or barriers etc.
- Corrosion test: The actual corrosion test is performed in 50 cycles. Each test cycle takes 24 hours and is divided in four phases:

    - Phase 1: Driving through wet and corrosive climate
    - Phase 2: Salt water spraying
    - Phase 3: Condensation at 45°C (113°F)
    - Phase 4: Incubation

This test creates within three months the corrosive behavior a vehicle under normal driving conditions would show at the end of its lifetime. Figure 7.87 compares a rear differential after the 50 days dynamic corrosion test (top) and after 7 years/ 200,000 km of regular lifetime (bottom).

Fig. 7.86 Dashboard before and after sunlight simulation (Source: BMW)

Fig. 7.87 Rear differential after 50 days dynamic corrosion test and after 7 years regular use (Source: BMW)

7.9.3.2 Simulation

Operational Strength

By use of kinematic and dynamic models, the stress mechanical components are exposed to in the overall system can be simulated. Simulation is done with realistic stress data that is recorded by driving specified test tracks with test vehicles that are equipped with load sensors. The load pattern so created then is used as input both for simulation and hardware testing (see below). The investigation of a virtual vehicle shown in Fig. 7.88 allows early detection of weaknesses in terms of operational strength.

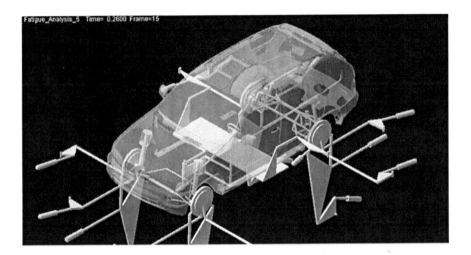

Fig. 7.88 Complete vehicle fatigue simulation (Source: BMW)

Aging

Numerical simulation methods for metallurgical and chemical processes such as aging or corrosion are an important current research topic; however, apart from material models for plastics that incorporate aging processes and sunlight simulation they are currently rarely applied in professional automotive development.

Embedded Software

Simulation of the hardware environment an embedded software program is supposed to control is a common way of early software testing. Here two different levels of simulation must be distinguished (compare Sect. 5.2.5):

- In a HIL simulation, both the sensors that create the input signals for an ECU and the actuators the ECU controls are emulated. This emulation represents the interface for the embedded system.
- In a SIL simulation, not only the relevant sensors and actuators, but also the ECU hardware are emulated. This approach allows evaluation of the software before the ECU is actually available.

7.9.3.3 Reliability Testing

Operational Strength

Generally, operational strength of mechanical components is validated through endurance tests. While the main focus here is on body, drivetrain and chassis, virtually any component that actively or passively moves undergoes continuous operational tests to prove the required reliability. For this purpose, a door e.g. is opened and shut 10,000 times.

The load pattern that has been recorded on real test drives as a realistic input for simulations (see above) is also used for dynamic whole vehicle test rigs. Extremely fast hydraulic servo valves controlling multiple hydraulic actuators allow the realistic replay of the recorded load data. Without leaving the development center, the vehicle under test can be exposed to the design relevant load spectrums, such as high speed circuit, moderate off-road driving, country roads etc. Figure 7.89 shows a vehicle undergoing such a fatigue test in a dynamic test rig.

Fig. 7.89 Complete vehicle fatigue test rig (Source: BMW)

Apart from validating the required operational strength, reliability testing also has to verify that the damage caused by exceptional stress levels or misuse such as hitting a curbstone or driving through potholes is kept within a reasonable limit. Figure 7.90 shows the front axle of a vehicle going through a pot-hole misuse test.

Fig. 7.90 Pot-hole misuse test (Source: BMW)

Embedded Software

After the formal tests of the program code as part of the software design process (see above), embedded software is tested for reliable functionality. In contrast to hardware, where final tests are expected to show the expected behavior and prove the assumptions made, software tests are carried out to find – and fix – as many "bugs" as possible. Figure 7.81 shows the general decrease of the failure rate of vehicle electronics during the burn-in phase towards the target value. It also shows the set-back the failure rate suffers after a functional up-date of the software.

As – even for relatively small programs such as in the ECU for a window lifter – testing all possible combinations of inputs is not feasible, it is important to deploy a test plan that specifies the test cases for checking the functionality specified in the requirements. For a comprehensive validation of the software's reliability, this test plan must also be assessed [54].

Electro-magnetic Immunity

As discussed above, immunity to electro-magnetical disturbance is an important aspect of complete vehicle reliability. After sensors, actuators, ECUs and harnesses have been designed according to EMC guidelines, the complete vehicle is tested for compliance. Immunity testing is carried out by exposing the vehicle to a specified spectrum of interference radiation in an anechoic chamber as e.g. specified by ISO/DIS 11451-2 as well as imposing electrostatic discharges according to ISO 10605. Test procedures are specified by or adapted to company standards. Figure 7.91 shows an anechoic chamber for complete vehicle testing. The antennae are used both for immunity testing and emission testing (see Sect. 7.10.5.3)

Fig. 7.91 Anechoic chamber for complete vehicle EMC validation (Source: BMW)

7.9.3.4 External Reliability Assessments

Just like passive safety or security from theft, reliability is a complete vehicle characteristic the customer can not directly experience in the show room or during a test drive. To ensure the vehicle in question for purchase meets their requirements in terms of reliability, customers seek the results of reliability assessments carried through by independent institutes (compare Sect. 6.4.3.3). JD Power's *vehicle dependability study* (VDS) e.g. evaluates vehicle quality after three years of ownership according to their owners' rating of problems they experienced over the previous 12 months. The ADAC's yearly breakdown statistics are based on a widespread assessment of vehicles that are between one and five years old and count the number of breakdowns per 1,000 vehicles registered. Figure 7.92 depicts the results from 2007.

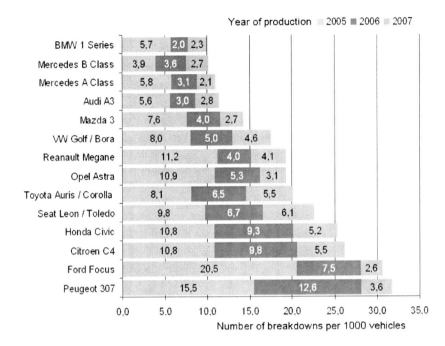

Fig. 7.92 ADAC 2007 breakdown statistics (Source: ADAC [48])

7.10 Sustainability

7.10.1 General Aspects

For a long time, sustainability in the automotive industry meant basically environmental friendliness, both of the cars and the respective production processes. For the carmakers, this environmental friendliness meant complying with the legal requirements and usually conflicted with their economic targets. When e.g. in the mid 1960s smog in cities like Los Angeles became unbearable, governments were pushed by the population and environmental activists to pass emission laws that obliged manufacturers to keep the amount of pollutants in their vehicles' emissions under specified limits in order to keep or restore local air quality. The same applies to fuel economy: In the early 1970s, during the first oil crisis, laws such as the *energy policy conservation act* in the U.S. were imposed to reduce fuel consumption.[27] The underlying motivation of course was more about the wish to become politically and financially more independent from foreign oil than the idea of saving the global reserves of fossil fuels.

Today, a much more integrated idea of sustainability is prevalent in the automotive industry. In addition to reducing or eliminating pollution and reducing consumption of exhaustible resources such as fossil fuels, sustainable development also embraces philanthropic and economic aspects. Maintaining an intact environment, a society in physical and psychic health, and economically successful business are the three main targets which are mutually interdependent and hence all of equal importance. In the long run, social responsibility, ecological reasonability, and economic management represent the most human and the most economic way of doing business. The term *eco-efficiency* denotes delivery of competitively priced goods and services that satisfy human needs and bring quality of life, while progressively reducing ecological impacts and consumption of resources throughout their life cycle, to a level at least in line with the earth's estimated carrying capacity [61].

Being a major driver for long-term economic success (and thus for long-term shareholder value) makes *corporate sustainability* also an important criterion for investors. Companies with a high degree of corporate sustainability embrace opportunities and manage the risks deriving from economic, environmental and social development. The *Dow Jones sustainability indexes* (DJSI) track the financial performance of the leading sustainability-driven companies worldwide and provides investors with a financial quantification of and a ranking in terms of sustainability.

---

[27] In Germany e.g., these laws went so far that for a period of time, private motor traffic was prohibited on Sundays.

Consequently, also another aspect of sustainability has changed over the years: Beyond the pure fulfillment of legal requirements, sustainability has become a customer relevant characteristic by itself. While formerly e.g. no automotive market (though some individuals) would have been willing to pay a price surcharge for the overfulfillment of legal emission requirements, customers today have started seeing sustainability as a distinct and important characteristic of a car and may be willing to pay for it. This trend becomes e.g. visible in the high demand for electric vehicles long before zero emission becomes mandatory (compare Sect. 7.10.2.4). Sustainability as an independent vehicle characteristic has four main aspects:

- Vehicle's consumption of fuel or other forms of energy
- Vehicle's emissions to the environment, including gases, noise and electromagnetic fields
- Treatment of vehicles at the end of their life
- Sustainability of the manufacturer's processes prior to vehicle delivery

Not least, the aspect of social acceptance also plays an ever more important role for motor vehicle customers. In broad areas of society, driving a car that is a notorious "gas-guzzler" or "polluter" brings along negative emotional reactions that can even turn into open aggression like the arson attack to a Hummer Dealership by environmentalists in Los Angeles in 2003. In an even stricter way, driving a vehicle produced by a manufacturer that is commonly associated with irresponsible management such as polluting production processes, unfair treatment of employees, or exploitation of international suppliers, is generally seen as unacceptable. As public and corporate management is especially expected to set a good example, meeting or exceeding expectations in terms of fostering sustainable mobility is even more important for cars that belong to corporate or public fleets.

Rating alternative mobility scenarios (such as the introduction of new energy concepts) concerning how well they represent sustainable development requires a sustainability assessment that fully covers four dimensions:

- The whole extended life cycle (marketing, research & development, production, distribution, usage, end-of-life treatment)
- The whole supply chain (including contractors, service providers etc.)
- All three aspects of sustainability (environmental impact, social impact, economic impact)
- All required products and services (vehicles, fuels, traffic environment)

Only consideration of all these aspects – as illustrated in Fig. 7.93 – allows objective rating of and subsequently deciding on alternative solutions.

		Process elements (incl. complete value chain)					
		Marketing	**R & D / planning**	**Production / realization**	**Distribution**	**Usage**	**End-of-life procedures**
System elements	**Vehicle**	Campaigns, events M. research	Design Testing	Manufacturing Logistics	Transport Sales	Driving Repair Maintenance	Dismantling Recycling Disposal
	Propulsion energy	Campaigns, events M. research	Experiments	Transformation from primary energy	Transport, Temporary conversion	Conversion to propulsion power	Disposal of residue
	Traffic environment	Campaigns, events M. research	Road planning Design of traffic control systems	Installation Road construction		Operation Repair Maintenance	Dismantling Recycling Disposal

Assessment criteria for each cell:
▶ Thoughtful utilization of exhaustible resources (e.g. water, fossil fuel ...)
▶ Minimum emissions and waste
▶ Work environments that sustains physical and psychic health
▶ Economic success

Fig. 7.93 Elements of a sustainability assessment

7.10.2 Energy Consumption and Tailpipe Emissions

As for most propulsion concepts energy consumption and related emissions are highly interdependent, both aspects of sustainability are summarized in this subsection.

7.10.2.1 Legal and Customer Requirements

Energy Consumption

Fossil fuels (diesel and gasoline) have been and still are the dominant source of energy for motorized vehicles. And for a long time, fuel economy has been seen as a matter of costs rather than a matter of economical and considerate exploitation of oil reserves (compare Sect. 7.2.1.2). The dramatic increase of oil prices over the last years however has brought the limited availability of worldwide oil supplies into the spotlight again. Long term world energy consumption scenarios like Shell's *Scramble* or *Blueprints* (see Fig. 7.94) make the need to use primary energy from renewable sources[28] rather than fossil fuels obvious and increase the

---

[28] Renewable energy sources are especially electricity created through solar, wind or hydro power, hydrogen, biomass or biofuels.

customers' willingness to make sustainability one of the criteria for their purchase decision.

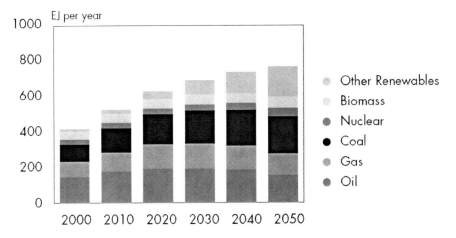

Fig. 7.94 Shell energy scenario *Blueprints* (Source: Shell [62])

As there are still no direct legal limits for fuel consumption, making fuel-inefficient vehicles costlier is the main governmental lever for fostering the reduction of fuel consumption. And worldwide, legislation has developed different means to charge both customers and manufacturers:

- Fuel costs as part of total vehicle costs have already been discussed in Sect. 7.2.1.2. A tax that puts the financial burden directly on the amount of consumed energy is a fuel tax. In Germany e.g., 68% of the costs filling stations charge for regular gasoline are taxes. The balance is simple: The more fuel a vehicle needs, the higher the costs for the user. Critics of this approach however call it inequitable, because it makes individual travel and the resulting pollution a privilege of those who can afford the fuel.
- In contrast to this, a so-called "gas-guzzler tax" is imposed to the customer on top of the purchase price of cars that fail to meet a minimum fuel economy level (e.g. 22.5 mpg in the U.S.). The obvious disadvantage is the "flat rate effect": After a customer has paid the one-time surcharge, the relative costs per mile driven becomes lower, the more miles the car is driven, which in turn motivates them to drive as much as possible – and thus to consume as much fuel as possible.
- Another governmental approach is to discourage production and sales of fuel inefficient vehicles by imposing taxes on manufacturers based on the average fuel consumption of the vehicles produced. An example for this is the penalty that manufacturers have to pay if the average fuel economy of their cars and light trucks sold in the U.S. exceeds *corporate average fuel economy* (CAFE) standards.

On the other hand, tax reductions or benefits like the permission to use car-pool lanes increase the attractiveness of fuel-efficient vehicles to potential customers.

Tailpipe Emissions

The first thing that comes into one's mind when talking about automotive emissions, are the substances in exhaust gases that result from the fuel combustion process:

- Carbon monoxide (CO)
- Carbon dioxide (CO_2)
- Hydrocarbons (C_nH_m), e.g. methane
- Nitrogen oxides (NO_X)
- Partly unburnt fuel
- Particular matter (PM),e.g. black carbon

Worldwide, legal limits for tailpipe emissions have dramatically tightened in recent years. The two main sets of regulations are the European EURO emission standards on one side, and the two U.S. standards issued by the *Environmental Protection Agency* (EPA) and the *California Air Resources Board* (CARB) on the other side.

European legislation has defined six different emission levels (EURO 1, EURO 2 ...), each representing a tighter maximum emission limit. Each EURO level is connected to a date at which the respective limit becomes legally required for all newly registered vehicles. The limits each standard provides differ depending on the fuel type (gasoline/diesel) and vehicle type (passenger car, light commercial vehicles etc.). Taking passenger cars with diesel engines as an example, the increasing requirements per standard are shown in Fig. 7.95.

In Germany, vehicles must be labeled with a windscreen sticker that indicates the respective standard they fulfill in terms of PM emission (Fig. 7.96 left). PM emission class 1 vehicles (e.g. gasoline engine vehicles without catalytic converter) do not get a sticker. Certain communities already refuse access to vehicles with low PM emission from their city centers. Figure 7.96 right shows the respective traffic signs that are displayed at the entrance of so called environmental zones and the allowed PM emission levels.

Currently, EURO standards do not limit CO_2 emission. However, a maximum average CO_2 emission of 120 g/km for all new passenger cars has been set as a target for the year 2012. The EURO standards have also been adopted by many other main markets such as China, India and Russia.

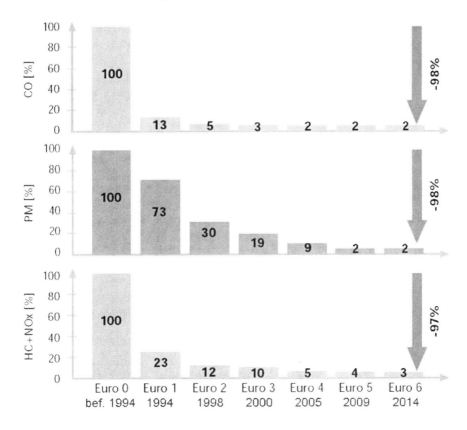

Fig. 7.95 Relative reduction of emissions over Euro standard stages (Source: VDA [63])

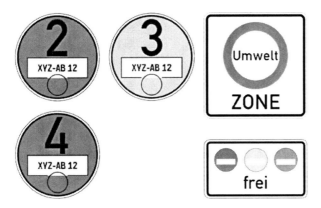

Fig. 7.96 German PM windscreen stickers and environmental zone
traffic sign (Source: Internet)

In the U.S., vehicles must either comply with federal EPA emission standards or – depending on the state the vehicle shall be registered in – with both the EPA and the much stricter LEV II emission standards set by CARB.

The currently valid EPA Tier 2 standard (2004–2009) allocates 10 different bins, according to the vehicle's emissions: Bin 1 defines the standard for the cleanest vehicle, bin 10 the least-clean.

All new vehicles sold in California, the so-called CARB-states (Maine, Massachusetts, New York, Vermont), or Canada must be certified according to CARB regulations. In the current phase 2, six different emission ratings are defined (see Table 7.14. A vehicle's emissions rating is posted on the *vehicle emission control information label* under the engine hood [64]. Apart from the actual maximum emission levels, PZEV and higher standards require long term functionality of emission control systems. In states other than the ones mentioned before, the less stringent EPA standards apply.

Table 7.14 CARB phase 2 vehicle emission ratings (Source: CARB [64])

Rating	Requirements
LEV II (Low Emission Vehicle)	NO_x emissions are 25% the level of a MY 1999 vehicle (LEV I)
ULEV II (Ultra Low Emission Vehicle)	HC and CO emissions are 50% lower than LEV II
SULEV II (Supra Ultra Low Emissions Vehicle	90% lower NO_x emissions and advanced emission control systems
PZEV (Partial Zero Emission Vehicle)	Meets SULEV tailpipe emission standards, has a 15-year / 150,000 mile warranty and has zero evaporative emissions
AT PZEV (Advanced Technology PZEV)	Meets SULEV tailpipe emission standards, has a 15-year / 150,000 mile warranty, has zero evaporative emissions and includes advanced technology components such as hybrid etc.
ZEV (Zero Emission Vehicle)	Zero tailpipe emissions and 90% cleaner than the average new model year vehicle

To make a new vehicle's rating in terms of fuel consumption directly visible to potential customers and thus facilitate comparison, the EPA *fuel economy label* (see Fig. 7.97) is mandatory in the U.S. since 2007. Concerning the vehicle's contribution to air pollution, the state of California will also introduce the *environmental performance label* (EP label) as depicted in Fig. 7.98. From model year 2009 on, new cars are required to display this sticker that shows the vehicle's scores regarding global warming and smog.

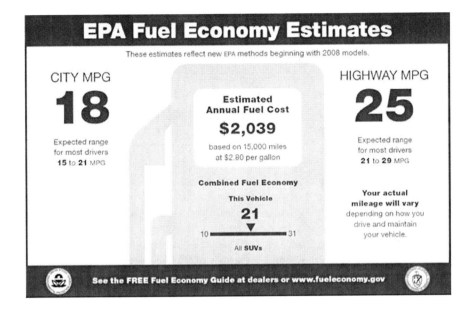

Fig. 7.97 EPA fuel economy label (Source: EPA [65])

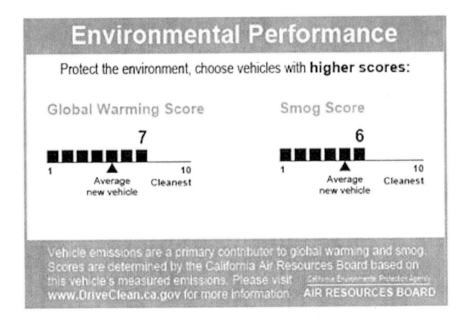

Fig. 7.98 CARB EP label (Source: CARB [66])

Other Requirements

In addition to setting limits to emissions, both European and U.S. standards require on-board diagnostic systems to check and ensure the functionality of emission reducing processes such as catalytic conversion, exhaust gas recirculation, or evaporation emission control systems. Also, the standards specify requirements concerning the quality of fuel, especially limits to the amount of sulfur.

Future Legal Requirements

In no other area of complete vehicle characteristics does legislation pose such a technical and economic challenge to carmakers as in vehicle emissions. To be able to do the R&D required to continuously fulfill the legal requirements, carmakers need to be exactly aware of coming changes in legislation. Figure 7.99 gives an overview of approved vehicle emission related legislation for the U.S., Canada, the European Union and Japan and a prognosis of its future development:

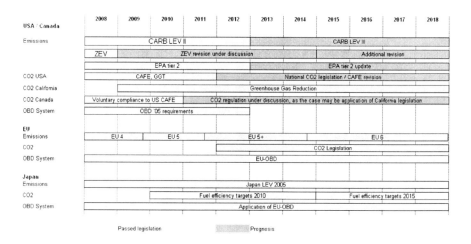

Fig. 7.99 Current and expected international legislation on motor vehicle emissions

7.10.2.2 Component and System Design

With both fuel consumption and tailpipe emissions being caused by the combustion of fossil fuel, most conceptual measures for efficiency and design for absence of emissions serve both targets.

Today, design for fuel-efficiency is mainly related to the introduction of advanced and alternative technologies such as electric engines, fuel cells, or biofuels. But with individual mobility being a necessity in most parts of the world rather than only convenience, the long-term move towards alternative propulsion concepts must also respect today's reality in terms of motorized traffic and the related industry:

- The worldwide fleet of passenger cars that is mainly propelled by combustion engines
- The automotive industry that has its main resources deployed to produce vehicles with combustion engines
- The oil industry that runs a worldwide infrastructure to supply individual vehicles with the necessary fuel

Even with automotive research and development running in high gear to make alternative concepts available for series production, combustion engines are – and for the near future will continue to be – the main propulsion technology for passenger cars. Hence, today's control levers of automotive fuel-efficiency comprise both the optimization of conventional solutions and implementation of new technologies (see Fig. 7.100).

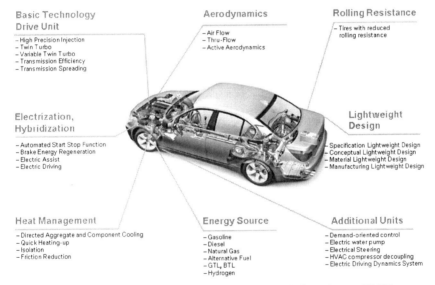

Fig. 7.100 Control levers of automotive fuel efficiency and emissions (Source: BMW)

7.10.2.3 Optimization of Internal Combustion Engines

Although continuously developed and improved for almost 150 years, internal combustion engines as the prevailing engine concept, still offer potential for improvement in terms of fuel efficiency and emissions. Such optimization can concern both the engine and the fuel technology.

Optimization of Engine Technology

Optimizing the core engine and its periphery can significantly reduce fuel consumption and related CO_2 emissions. As an advantage, this approach does not require changing an existing vehicle architecture or energy supply infrastructure.

Taking four consecutive model years of the BMW 530i sedan as an example, Fig. 7.101 demonstrates how fuel consumption could be reduced by introducing the following features:

- VANOS (variable valve timing)
- Valvetronic (continuously variable intake valve lift)
- High precision injection
- Weight reduction (e.g. Mg/Al composite crankcase)

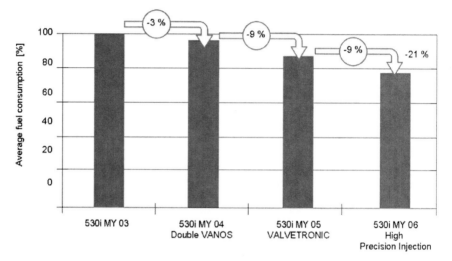

Fig. 7.101 Features reducing fuel consumption (Source: BMW)

Combustion of Renewable Fuels

A way to become independent from exhaustible reserves of fossil fuels (rather than reducing energy consumption or emissions) is by running internal combustion engines on biofuels – solid, liquid or gaseous fuels derived from (recently) dead biological material. Biofuels that can be directly used in internal combustion engines are:

- Bioethanol and other bioalcohols, produced by microorganisms and enzymes through the fermentation of sugars or starches, or cellulose
- Biogas, generated through anaerobic digestion of biodegradable materials such as biomass, manure, sewage, food waste, or dedicated energy crops
- Biodiesel, generated by transesterification of vegetable or animal fat

A *well-to-wheel analysis* that summarizes the ecological impact of production, distribution and consumption however shows that production (including the usage of fertilizers and herbicides) and transport of some alternative fuels require more fossil resources and cause more greenhouse gas emissions than the equivalent amount of fossil fuel. The amount of CO_2 e.g. that is created during coal liquefaction is twice as high as the amount created by combustion of the same amount of conventional gasoline. Another example is the *Texas Commission on Environmental Quality's* (TCEQ) planning to ban B20 – a mix of biodiesel and petroleum based diesel – due to its high NO_x emissions. An additional negative side effect of biofuels is the increase of food prices through farming land, resources and equipment used for cultivating and harvesting energy crops rather than food crops ("food vs. fuel" dilemma).

Hybridization

The energy efficiency of vehicles equipped with conventional internal combustion engines is fairly poor – to say the least. As Fig. 7.102 shows, over 80% of the energy incorporated in the fuel is wasted as heat transferred to the cooling system or through the exhaust gas.

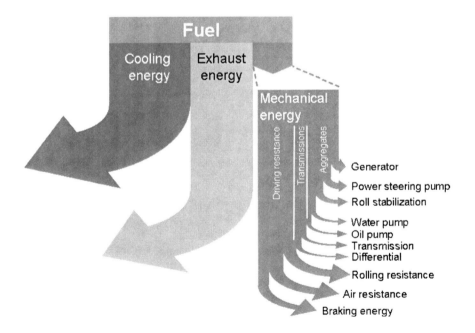

Fig. 7.102 Utilization of fuel energy in combustion engine vehicles (Source: BMW)

Based on this, further efficiency improvement of motor vehicles requires an advanced energy management approach that reuses as much energy as possible. This requires both in-depth understanding of the energy flows in the vehicle (see Fig. 7.103) as well as technologies that allow the transformation and storage of wasted energy for later usage.

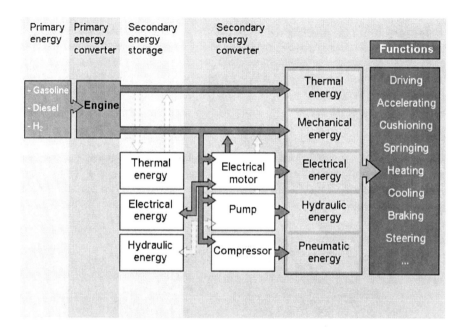

Fig. 7.103 Inner-vehicle energy flow (Source: BMW)

The main elements of energy management are heat management, brake energy recuperation, and electrification:

- Heat management includes measures to reduce friction, keep the engine from cooling off when the engine is switched off, or transform heat into electric energy by means of thermoelectric elements.
- By means of brake energy recuperation, a part of the kinetic energy that is released during deceleration is converted into electrical energy by means of a generator and stored in the electrical energy storage.
- Auxiliary units (such as HVAC compressors) which are usually permanently engaged in the engine's belt drive are actuated by electric motors that only consume energy when needed.

An established concept of energy management are *hybrid electrical vehicles* (HEV) which recuperate brake energy, switch off the engine when not needed, or use an electric engine to temporarily boost or permanently drive the vehicle. Figure 7.104 compares the different levels of hybridization.

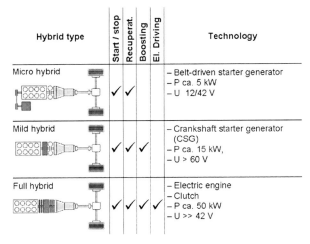

Hybrid type	Start / stop	Recuperat.	Boosting	El. Driving	Technology
Micro hybrid	✓	✓			– Belt-driven starter generator – P ca. 5 kW – U 12/42 V
Mild hybrid	✓	✓	✓		– Crankshaft starter generator (CSG) – P ca. 15 kW, – U > 60 V
Full hybrid	✓	✓	✓	✓	– Electric engine – Clutch – P ca. 50 kW – U >> 42 V

Fig. 7.104 Levels of hybridization

Especially in inner-city driving with lots of stop-and-go, hybrid vehicles show a significant improvement in fuel efficiency. Figure 7.105 demonstrates the efficiency contributions of automatic start/stop, recuperation, and electrical driving over a part of the *New European Driving Cycle* (NEDC), compare Sect. 7.10.2.5).

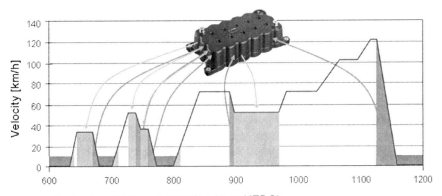

Reduction of fuel consumption (over NEDC):
- Automatic start stop system: ~ 5%
- Brake energy recuperation: ~ 5%
- Electric driving on stored energy: ~ 5%

Fig. 7.105 Hybridization efficiency contributions (Source: BMW)

Among the disadvantages of hybrid systems are the additional costs and the additional weight stemming from the additional components. In some cases, the additional weight partially neutralizes the efficiency effects gained by the hybridization. Considering the long-term goal of providing zero emission vehicles, it is clear that hybrid powertrains can only be a temporary solution.

7.10.2.4 Alternative Powertrain Concepts

While the optimization of existing engine concepts and hybridization can bring significant improvements, the long-term goals of zero emission vehicles, renewable primary energy, and energy security require alternative powertrain concepts. Today, there are mainly two engine types that allow emission-free conversion of stored energy in propulsion: Hydrogen fueled combustion engines, and electric motors. Based on these two alternatives, there are three competing concepts for ZEVs which are discussed below:

- Hydrogen vehicle: Hydrogen fueled combustion engine with hydrogen stored on-board
- Fully electric vehicle: Electric motor with electric energy stored on-board
- Fuel-cell vehicle: Electric motor with electric energy generated on-board from hydrogen by a fuel cell

Rating and comparing these alternative concepts in terms of sustainability however may not be done by measuring fuel efficiency and emissions of the vehicle itself; it rather requires a realistic life cycle assessment that summarizes their overall ecological impact, including energy consumed and emissions caused during:

- Development (e.g. emissions caused during hydrogen tank explosion tests)
- Marketing (e.g. electric car races as marketing events)
- Manufacturing (e.g. energy required to manufacture a battery pack)
- Transport (e.g. emissions caused by an airplane that ships electric motors from an overseas supplier)
- Decommissioning (e.g. energy required to neutralize vehicle fluids at end-of-life)

With respect to energy consumption, the relevant question is – independent from the lifecycle phase – whether the primary energy source is renewable or not. Renewable sources for electricity e.g. include solar power, wind power, and hydro power.

Hydrogen Vehicles

The clear advantage of hydrogen compared to fossil fuel or biofuels lies in its zero-emission capability: The combustion of hydrogen generates high amounts of thermal energy and only pure water as a byproduct. Furthermore, hydrogen can be used with existing combustion engine concepts. So-called bifueled engines are equipped both with a conventional injection system and a hydrogen-air-mixing-device, and thus can be smoothly switched back and forth between gasoline and hydrogen mode. This technology enables vehicles to utilize the emission-related benefits of hydrogen fuel while at the same maintaining the independent mobility that is assured by the existing gasoline distribution net. As an example, Table 7.15 summarizes the technical data of the bifueled V12 hydrogen engine that propels the BMW Hydrogen 7.

Table 7.15 Technical data of V12hydrogen engine (Source: BMW)

Engine type	V12
Fuel type	Bifueled: LH2 (liquid hydrogen) and gasoline
Displacement	6.0 liters
Max. output	191 kW (260 bhp) at 5100 rpm
Max. torque	390 Nm at 4300 rpm
H2 Carburetion	Direct injection
Gasoline carburetion	External hydrogen-air-mixture

However, despitethe benefits of hydrogen engines technical challenges remain concerning safety, cost and efficiency of the production and handling processes that must be solved before hydrogen can become an alternative that is suitable for mass motor traffic. Concrete requirements are:

- Sustainable and affordable production of hydrogen
- Safe, sustainable and affordable distribution of hydrogen
- Safe on-board storage of hydrogen

Today, hydrogen is almost exclusively produced by steam reforming of natural gas – a process that actually converts the major part of the natural gas' energy into CO_2 that is released into the environment. From the viewpoint of sustainability, electrolysis (especially high-temperature electrolysis) of water and biological production by algae are two promising alternative concepts of hydrogen production. With research and industrialization advancing, efficiency and costs for these processes can be expected to improve.

Distribution of hydrogen to individual motorists requires a new infrastructure of central or distributed production facilities, pipelines, storage containers, transporters and distribution facilities (filling stations). As hydrogen can ignite with a very low energy input, diffuses through metals, and additionally embrittles steel, safety is a major concern for transportation and temporary storage.

The biggest issue in using hydrogen fuel in passenger cars is safety of on-board storage. The amount of hydrogen required to allow a sufficient range can only be stored in a reasonable package volume if in liquid form. This however requires permanent cryogenic storage at temperatures below –250°C to prevent boil-off. Figure 7.106 shows such a cryogenic tank system for approx. 8 kg or 170 l of liquid hydrogen. If the cooling system should fail, the high vacuum insulation with multiple layers of reflective aluminum foil keeps the liquid hydrogen from boiling off for about 17 h. Due to the high flammability of hydrogen mentioned above, passive safety requirements for vehicles with on-board hydrogen storage are extremely high.

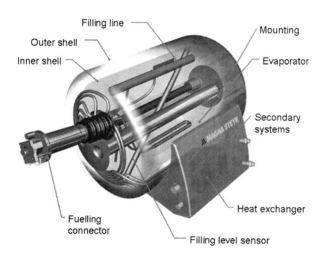

Fig. 7.106 Tank for cryogenic on-board storage of liquid hydrogen
(Source: MAGNA Steyr)

Based on the BMW 7 Series, the Hydrogen 7 is the first hydrogen propelled vehicle that was developed over a realistic PEP (compare Chap. 1), thus proving not only the technical feasibility of the product concept, but also demonstrating that this concept can be developed, manufactured and sold in an existing business environment. Figure 7.107 shows the main elements of the Hydrogen 7's powertrain:

In hydrogen mode, the Hydrogen 7 accelerates from 0 to 100 km/h (62 mph) in 9.5 sec, the maximum speed is electronically limited to 230 km/h (143 mph). The total range is 700 km (435 miles), 200 km (124 miles) on hydrogen and 500 km (311 miles) on gasoline.

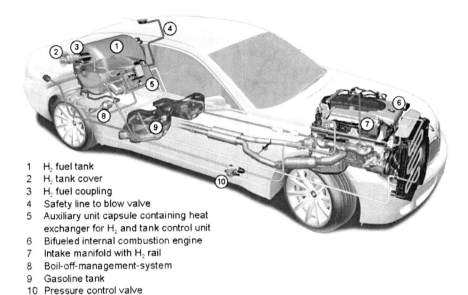

1   H₂ fuel tank
2   H₂ tank cover
3   H₂ fuel coupling
4   Safety line to blow valve
5   Auxiliary unit capsule containing heat
     exchanger for H₂ and tank control unit
6   Bifueled internal combustion engine
7   Intake manifold with H₂ rail
8   Boil-off-management-system
9   Gasoline tank
10 Pressure control valve

Fig. 7.107 Layout of major components in a hydrogen vehicle (Source: BMW)

Fully Electric Vehicles

Although the first vehicles with electric engines had already been built during the first half of the 19th century, it was the growing awareness for fuel consumption and emissions that brought electric drive vehicles back into the showrooms during the 1990s. Compared to combustion engines, electric motors have clear advantages:

- Conversion of electrical energy into propulsion is 100% emission free
- Higher level of efficiency (~ 95%)
- Higher energy density
- Maximum torque available starting from 0 rpm
- Lower exterior noise level
- Simple drivetrain concepts
- No cooling necessary

But while electric motors show superior characteristics, the drawback of electric drive concepts currently is on-board storage of electrical energy. Compared to liquid or gaseous fuel in a tank, battery packs available today have the following disadvantages:

- High costs
- Short lifetime

- Restrictions in package, weight, performance and/or range through low energy density
- Inconvenient "refueling" through long charging time

The following example illustrates the challenge: A fuel tank filled with 60.0 l (15.9 gal) of fuel represents 1.88 GJ of energy stored on board, weighing 45.0 kg (99.2 lb) without the tank. With an average fuel pump delivering 50 l/min (13.2 gal./min), the tank can be completely filled within 1 min 12 sec. An advanced Li-ion battery pack that can store the same amount of energy would weigh about 4430 kg (9766 lb) and need about 2600 l of package space. Using a 10 kW outlet, full charge of the battery pack would require 52 h.

Research and industry however have put tremendous effort into improving battery concepts (e.g. Li-polymer) or finding alternative ways to store electric energy, such as super and ultra capacitors or flywheel systems.

An impressive sports car that represents the capabilities of electrical vehicles today is the Tesla Roadster which has been on sale since 2008. Table 7.16 shows technical data for the powertrain, battery system and complete vehicle:

Table 7.16 Technical data of the Tesla Roadster (Source: TESLA Motors [67])

Powertrain	Engine type	375 volt AC induction air-cooled electric motor
	Max net power	185 kW (248 HP) @ 4500-8500 rpm
	Max rpm	13,000 rpm
	Max torque	375 Nm (276 ft/lb) @ 0-4500 rpm
	Efficiency	92% average, 85% at peak power
	Transmission	Single speed fixed gear (drive ratio 8.28:1)
Battery system	Type	6,831 Li-ion cells
	Capacity	191 MJ
	Charging time	210 min
	Weight	450 kg (992 lbs)
	Estimated lifetime	160.000 km (100.000 miles)
	Price	~20,000 US$
Performance	Acceleration	0 to 100 km/h (0 to 60 mph): 3.9 s
	Top speed	201 km/h (125 mph) electronically limited
	Range:	356 km (221 miles) on the EPA combined cycle
	Weight	1,238 kg (2,723 lb)

While tailpipe emissions of electric vehicles truly are zero, generation of required electrical energy might significantly contribute to emissions of CO_2 and other greenhouse gases if e.g. the source of grid electricity is coal or other fossil fuels.

Fuel Cell Electric Vehicles

In contrast to combustion engines, fuel cells convert hydrogen into electric energy which then is used to propel the vehicle by means of an electric motor. Fuel cells in combination with electric motors are also 100% emission free and in addition have a higher degree of efficiency than combustion engines: In the NEDC, the tank-to-wheel efficiency of a *fuel cell electric vehicle* (FCEV) is about 36% while the corresponding value for a diesel engine is 22% [68]. As fuel cells have no moving parts, they are also very reliable. Compared to combustion engines however, fuel cells are expensive: Production of a fuel cell consumes about 2.5 times more energy than it creates over its whole life. As fuel cells can not run on regular gasoline or diesel fuel, their usage is restricted by the availability of hydrogen – or require additional batteries to store electrical energy. With the technology still not ready for mass production, passenger cars with hydrogen fuel cells like e.g. the Honda FCX Clarity can not be purchased but are leased to selected and pre-qualified customers.

7.10.2.5 System Integration and Validation

Both energy consumption and propulsion related emissions highly depend on whether the vehicle is driven in inner-city traffic, on extra-urban roads or on inter-states, and on the actual manner of driving. To be able to create comparable values – especially in emission testing, vehicles are driven according to standardized driving cycles which specify a certain speed pattern for a given period of time.

Driving cycles are either modal or transient. Modal cycles – preferred in Europe and Japan - are rather artificial, with distinct phases of linear acceleration, linear deceleration, and constant speed, which allow simple calculations. Transient cycles – preferred in the U.S. – are derived from real driving patterns and thus represent a more realistic driving behavior than modal cycles do.

In Europe, the most important driving cycle is the modal NEDC (see Fig. 7.108), which is also called the ECE cycle. Specified by Directive 70/220/EEC, it consists of a series of four ECE 15 city cycles with acceleration and deceleration maneuvers at a maximum speed of 50 km/h (30 mph), and one EUDC overland cycle with a maximum speed of 120 km/h (70 mph). The NEDC cycle is 11 km (6 miles) long and takes 20 min [69].

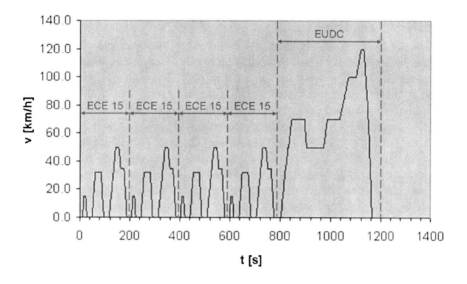

Fig. 7.108 New European Driving Cycle (Source: 70/220/EWG [69])

Due to the disadvantages of modal driving cycles especially for the assessment of hybrid vehicles, the transient HYZEM cycle (see Fig. 7.109) was developed as part of a joint European research project. HYZEM consists of three parts: 600 s urban traffic, 800 s extra-urban traffic, and 1806 s highway traffic.

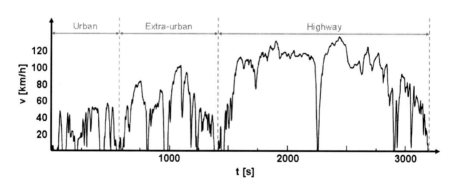

Fig. 7.109 HYZEM driving cycle (Source: Brite-Euram [70])

For the U.S., the EPA has defined four different driving cycles: FTP 75 (see Fig. 7.110) for city driving at low speeds in stop-and-go urban traffic (also used for a cold temperature test), the *highway fuel economy cycle* for free-flowing traffic at highway speeds, the *US 06 Supplemental FTP cycle* (so-called "*aggressive driving cycle*") with high engine loads, and the SC03 *speed correction cycle* for a test that includes air conditioner use under hot ambient conditions.

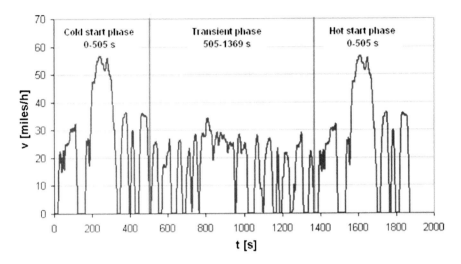

Fig. 7.110 EPA FTP 75 driving cycle (Source: EPA)

In Japan, two stylistic driving cycles have been established: The *10 mode* cycle represents 135 s of urban driving at a maximum speed of 40 km/h. The *15 mode* cycle is a mixture of urban and extra-urban driving with a maximum speed of 70 km/h and lasts for 220 s. *10-15 mode* is a sequence of one 15 mode cycle, three 10 mode cycles and one 15 mode cycle.

Figure 7.111 compares the driving cycles mentioned above in terms of average speed and relative positive acceleration.

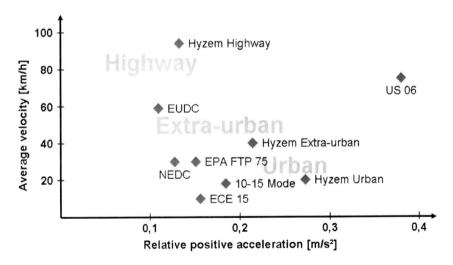

Fig. 7.111 Comparison of international driving cycles

Emissions and energy consumption are tested on dynamometers (see Fig. 7.112 and 7.113), where the vehicle exactly follows the speed pattern of the respective driving cycle.

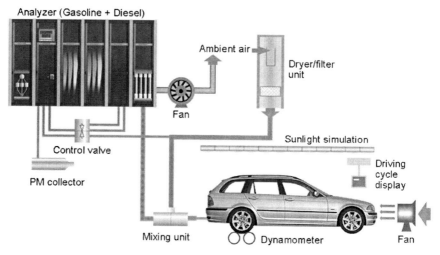

Fig. 7.112 Emissions testing on dynamometer (Source: Pierburg Instruments)

Fig. 7.113 Emissions testing on dynamometer (Source: BMW)

To make the results as precise as possible, both the vehicle and the test environment must adhere to exactly specified conditions (e.g. ambient temperature, engine temperature, battery state of charge, fuel tank level, auxiliary ventilation temperature and air flow etc.). Tailpipe emissions are then collected over the respective driving cycle, mixed with cleaned ambient air and analyzed.

7.10.3 Evaporative Emissions

7.10.3.1 Legal and Customer Requirements

In addition to the byproducts of the combustion process (or other forms of converting energy into propulsion power), a vehicle emits *volatile organic compounds* (VOCs) both into the cabin and the environment – even with the engine switched off. The two main types of such emissions are evaporations from the fuel system and evaporation from plastic components.

Fuel[29] evaporating out of the fuel system (tank, lines, carburetor) or during the filling process represents a major share of a vehicle's total hydrocarbon pollution. Especially at high ambient temperatures, the amount of evaporated hydrocarbon can actually exceed the amount that is released with the exhaust gas. The three types of evaporative fuel losses are:

- Diurnals: Fuel evaporating due to increasing ambient temperature
- Hot soaks: Fuel evaporating immediately following vehicle use caused by heat emission from the engine
- Running losses: Fuel evaporating while the vehicle is in operation (so-called "crankcase emissions"

Non-fuel evaporative emission (also called "resting losses") are primarily comprised of VOCs that evaporate from interior and exterior plastic components (especially softeners and solvents), non-fuel liquids (such as lubricants, brake fluid, or screenwash), or adhesives. As the mass of evaporated VOCs increases with the surface area of the emitting plastic part, foams are especially critical.

Legal requirements concerning the limitation of evaporative emissions of motor vehicles have dramatically tightened over the last years. While UNECE R34 / EURO 1 (compare Sect. 7.10.2.1) in 1993 allowed 20 g of evaporative emissions over a 24 h cycle, EURO 4 states since 2007 that in the same period of time emissions should not surpass 2 g. For EURO 5, which will be introduced in 2009, the limit of 2 g/24 h will probably remain but will be also binding if fuels are used that contain 5% ethanol and thus have a higher vapor pressure. In the U.S., CARB

---

[29] As Diesel fuel evaporates at a much higher temperature, this mainly applies to gasoline.

LEV II vehicles (compare Table 7.14) may have a maximum diurnal emission of 500 mg, a PZEV only 350 mg of which a maximum 54 mg my come from the fuel system. Taking into account that a typical passenger car weighs about 1.6 t and contains about 8,000 plastic parts, 300 mg of non-fuel evaporations can hardly be traced.

7.10.3.2 Component and System Design

Fuel Evaporation

Minimization of evaporating fuel (diurnals, hot soaks and running losses) escaping to the atmosphere is achieved today by evaporative control systems, which catch and store fuel evaporating from the fuel system in a charcoal canister. As soon as the engine is started, the stored fuel vapors are sucked out of the canister and fed via a purge valve to the engine where it is burnt as part of the fuel-air-mixture. In addition to the canister and the purge valve, the complete vehicle package must provide space for a fuel tank that is not only impermeable to fuel vapors but also has additional volume to collect the fuel vapors. Just as for other emission reducing measures, legislation requires a long term warranty for such fuel evaporation control systems.

Fuel evaporation depends largely on the fuel's vapor pressure. In order to minimize fuel evaporation, the fuel vapor pressure must be as low as possible; on the other hand, the vapor pressure must still be sufficiently high to safely allow cold starts. Thus, refineries adjust the vapor pressure to the climactic conditions during the year.

Non-fuel Evaporative Emissions

In a premium vehicle today, about 50% of the non-fuel evaporative emissions (resting losses) originate from chassis parts (especially tires), 25% from exterior and interior trim, and 25% from plastic engine components. The vast reduction in HC evaporation from plastic parts over the last years has been achieved in large part due to the improvement of polymers such as polyoxymethylene (POM) or poly-para-phenylene (PPP). While 10 years ago e.g. 15 POM clips could have caused the same amount of emissions as the whole interior, their single contribution today can hardly be measured. In addition to material improvements, parts with a relatively high emission level (e.g. seat or carpet foams) can by isolated by covering or coating them with less emitting material. A tight leather or fabric cover of a seat foam part e.g. encapsulates the hydrocarbons evaporating from the foam and prevents it from being released to the environment.

With fuel evaporation being significantly reduced, emissions from lubricant oils and other vehicle fluids have now come into the center of attention. Tightness of canisters, pipes and filler caps are of high importance. A small amount of windscreen washing fluid spilled out of the container can immediately increase the non-fuel evaporative emissions tenfold.

7.10.3.3 System Integration and Validation

EURO, EPA, and CARB regulations require measurement of the total evaporative emissions (diurnals, hot soaks, running losses, resting losses) released by the complete vehicle. Even though these emissions are caused by distinct parts or liquids, the total amount of evaporation usually differs significantly from what the sum of the individual part's emission would be. The reason for this effect is chemical reactions between the different elementary emissions that can decrease or increase total emission by either mutual neutralization or creation of additional HC emissions. The underlying chemical processes are however not fully understood so that it is always necessary to measure the complete vehicle.

Complete vehicle evaporative emissions are measured through *sealed housing evaporative determination* (SHED) tests which are specified e.g. by EU directive 98/69/EC / UNECE R83, CARB and EPA 40 CFR 86.133-96. In a SHED test, the vehicle is put in a gas tight test chamber (see Fig. 7.114) at specified temperature, humidity and air flow for a specified period of time.

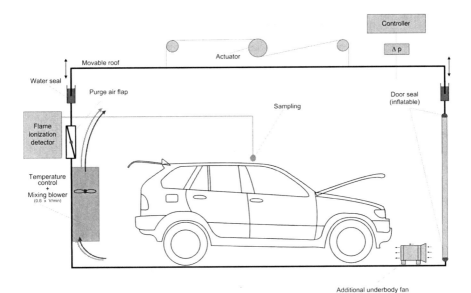

Fig. 7.114 Complete vehicle SHED test chamber (Source: BMW)

The amount of hydrocarbons released during the test period is measured at the end of the test by a flame ionization detector. For the measurement of running losses, the test is performed with the vehicle running on a dynamometer within the test chamber. In this case, the exhaust gases are routed out of the chamber so that only the evaporative emissions remain and are measured.

For pre-qualification of new components and for problem analysis, SHED tests are also carried out with individual parts and components. Test chambers for component tests are much smaller (e.g. 1 m³) and are specified e.g. by VDA 276 or SAE J1737.

Rating and comparison of different plastic materials concerning their evaporative behavior can be done by enclosing the respective resin in a flask which is exposed to temperatures between 80 and 100°C and then statically (VDA 277) or dynamically (VDA 278) analyzing the trapped gasses (so-called "headspace").

7.10.4 Noise Emissions

7.10.4.1 Legal and Customer Requirements

Almost no vehicle characteristic is perceived as ambivalently as exterior vehicle noise. What the customer of especially sporty cars might see as a wonderful sound that suits a powerful engine or vehicle (and might even amplify e.g. by means of special exhaust pipes) is usually perceived as a pure annoyance by the public. Noise demonstrably has detrimental effects both on human (and animal) psyche and physis, and traffic noise made up from road, rail and air traffic is the biggest source of noise pollution. For the overall noise emission of motorized road traffic, the three determining factors are:

- Vehicles: Engine, exhaust system, tires, aerodynamics (see Sect. 7.4.2 for sources of vehicle noise)
- Traffic parameters: Density, max. allowed speed, traffic flow constancy
- Roadways: Road surface, layout, ascents and descents

Current legal requirements concerning exterior noise only cover accelerated pass-by-noise and hence mainly limit the sound level of the exhaust system. Since 1995, UNECE regulation R51 requires a maximum pass-by-noise level of 74dB(A) for passenger cars, measured according to ISO 362 (see below) which is equivalent to SAE J1470. During constant-speed driving however, tire/road noise represents about 70% of the sound pressure, outweighing noise coming from the drivetrain, exhaust system or exhaust tip by far [71]. In Europe, rolling noise is limited by EU-directive 2001/43/EG, measured according to ISO 13325.

Especially for premium sedans, customers however expect their vehicle's sound emission to be well below the legal limits, and manufacturers put significant effort into noise reduction. On the other hand, vehicles with very low noise emission can actually represent a safety risk for other traffic participants: As e.g. electric cars can hardly be heard when driven at low speed, these traffic participants often risk not being perceived by pedestrians and bicyclists.

7.10.4.2 Component and System Design

Exterior vehicle noise is especially critical in typical suburban traffic, characterized by a large number of cars, frequent acceleration and deceleration, and low to medium speeds. Generally, measures to reduce noise emission caused by motorized traffic include [71]:

- Measures that reduce noise emission (relating technical characteristics of vehicles, technical characteristics of roads, traffic volume and traffic speed)
- Anti-propagation measures

From automotive design's point of view, noise reduction measures must relate to the components that mainly drive exterior noise (see Fig. 7.115). Subsequently, design measures to reduce exterior vehicle noise include:

- Optimization of muffler systems
- Optimization and encapsulation of engine and gearbox
- Optimization of tires

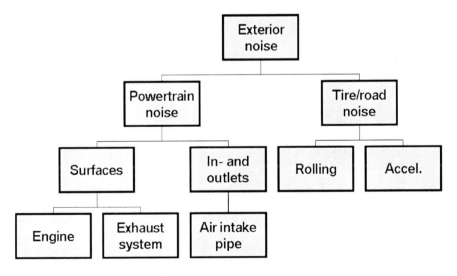

Fig. 7.115 Sources of exterior vehicle noise

A huge noise reduction potential – that however is not in the responsibility of the vehicle manufacturers – lies in the provision of an optimized traffic environment: Measures include road design that – together with tire design – optimizes tire-road-noise as the prevailing source of noise emission caused by suburban traffic, introduction of inner-city traffic flow control systems that reduce the need for vehicles to accelerate or decelerate, and absorbing noise barriers.

7.10.4.3 System Integration and Validation

The most important international test standard for measuring external vehicle noise is the pass-by test according to ISO 362 (equivalent to SAE J1470). In this test, the vehicle noise is measured by means of two microphones, one on either side of the vehicle (see Fig. 7.116). The vehicle under test drives into the testing route at a speed of 50 km/h. 10 m in front of the microphones, the vehicle then accelerates until the end of the vehicle is 10 m away from the microphones. Drivers usually use special tools that assist them in choosing the correct gear and entry speed.

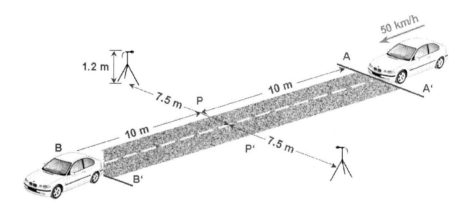

Fig. 7.116 Set-up for the ISO 362 pass-by noise test (Source: ISO [72])

To be able to separate engine and exhaust noise from tire/road noise, the ISO 362 test can also be carried through on a dynamometer. This setup then requires 36 pairs of microphones in a semi-free field. Figure 7.117 shows such an exterior noise test facility. An additional advantage of such facility is that tests can be carried through at standardized climate conditions.

Fig. 7.117 Exterior noise test facility (Source: BMW)

Other standardized test procedures that relate to vehicle noise emission are the coast-by test (tire noise test) according to ISO 13325, and the exhaust noise test according to ISO 5130.

7.10.5 Electro-magnetic Emissions

7.10.5.1 Legal and Customer Requirements

While in Sect. 7.9.2.1 EMC was discussed from the perspective of electro-magnetic immunity, a second and no less important aspect is electro-magnetic fields vehicles send out to their environment. Vehicle manufacturers must satisfy legislative EMC requirements concerning broadband and narrowband emissions measurements in the 30-MHz to 1,000-MHz band.

Concerning electro-magnetic emissions of their car, customers typically do not expect the vehicle to meet more than the legal requirements. Worldwide, legal limits for electro-magnetic emissions are set by UNECE regulation R10 (which implements EU-Council Directive 72/245/EEC, version 2004/104/EU).

7.10.5.2 Component and System Design

In general, reduction or of electro-magnetic emissions is achieved by the same design measures that are taken to obtain immunity from electro-magnetic fields. These measures are discussed in Sect. 7.9.2.1.

7.10.5.3 System Integration and Validation

As there are no official standards that completively specify the EMC test procedures, OEMs typically have brought their own standards into operation. BMW Group Standard GS 95002 [73] e.g. includes specifications for four types of test procedures for electro-magnetic emissions from components and complete vehicles:

- Galvanic measurement, including measurement of pulse width, frequencies, and slew rate of clocked signals.
- Capacitive measurement using a coupling clamp according to ISO 7637-3. In contrast to ISO7637-3, the coupling clamp is used for the recording of the emitted electric fields.
- Inductive measurement using a current probe and measurement receiver.
- Field coupled measurement using a stripline (derived from ISO 11452-5) or antennas in an anechoic chamber according to CISPR25.

7.10.6 Treatment of End-of-life Vehicles

7.10.6.1 Legal and Customer Requirements

Sustainable development generally requires processes that leave the environment at least in the condition it was before. Considering, that every year nearly 12 million vehicles are taken out of service only in North America, neutralization of the environmental impact of these *end-of-life vehicles* (ELV) is a huge challenge for the automotive industry. The related processes at the end of a vehicle's intended use serve two main targets:

- Isolation of toxic or harmful materials to prevent them from being released to the environment
- Re-utilization of the biggest possible portion of the vehicle's components and materials

A comprehensive package of legal requirements concerning treatment of end-of-life vehicles is offered by EC directive 2000/53/EC, the European end-of-life vehicles directive. Its main requirements are [74]:

- Extended producer responsibility principle: Manufacturers must take back vehicles once they have reached the end of their lives at no cost to the consumer. The cost aspect of end-of life vehicles has already been discussed in Sect. 7.2.1.1.
- Hazardous materials: The use of hazardous or toxic materials such as lead, chromium VI, cadmium and mercury in the vehicle production process is prohibited.
- End-of-life vehicle collection: An adequate collection system ensures transfer of end-of life-vehicles to legitimate treatment facilities at the cost of the producer. Correct removal of critical components such as batteries, tires, operating fluids, hazardous components, carbon fiber components or air bags are ensured by certified dismantlers, who also issue a *certificate of destruction* which is required for the mandatory de-registration by the last owner.
- Requirements on re-use and recovery: Vehicles put on the market after 2006 and 2015 must have recycling rates of 80 and 85%, respectively. To facilitate the dismantling process mentioned above, vehicle manufacturers must ensure consistent material-related coding of components. Disassembly manuals that describe the specific recycling process for the vehicle (including identification of hazardous components) must be distributed by the vehicle manufacturer six months after market entry at the latest.
- Mandatory customer information: Along with promotional material or sales catalogues, potential customers must be provided with information e.g. concerning the recyclability of the vehicle and its components, the treatment of end-of-life vehicles offered by the manufacturer.

Legal requirements in the U.S. are substantially lower than in Europe. An example for voluntary activity is the *End Of Life Vehicle Solutions Corporation* (ELVS), which was created by automotive OEMs[30] in the U.S. to promote the industry's environmental efforts in recyclability, education and communication, and the proper management of substances of concern. As part of EPA's *National Vehicle Mercury Switch Recovery Program* (NVMSRP), ELVS facilitates recycling of mercury used in older cars' lighting and ABS system switches in a close cooperation with public and private industry partners. By March 2008, the program has recovered one million switches containing over a ton of mercury, which otherwise would have been disposed of in landfills [75].

In addition to the legal requirements outlined above, each OEM has developed its own corporate standards and norms.

---

[30] Participating Members of ELVS are: BMW, Chrysler, Daimler, Ford, General Motors, Mack, Mitsubishi, Nissan, PACCAR, Porsche, Subaru, Toyota, Volkswagen, Volvo.

7.10.6.2 Component and System Design

A broad description of end-of-life procedures for motor vehicles together with methods to implement easy recycling and dismantling during the product development process is published as guideline VDI 2243 [76] by the *Association of German Engineers*. (VDI).

Motor Vehicle Dismantling Procedure

To create the technical prerequisites for proper end-of-life treatment, vehicle designers must consider each step of the motor vehicle's end-of-life procedure (see Fig. 7.118) and the respective design requirements from the concept phase on.

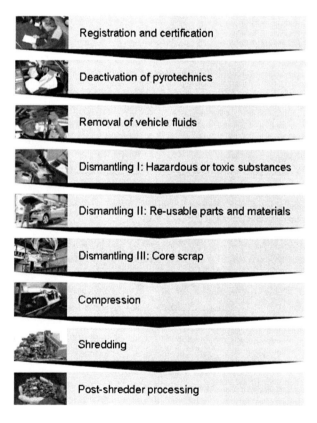

Registration and certification

Deactivation of pyrotechnics

Removal of vehicle fluids

Dismantling I: Hazardous or toxic substances

Dismantling II: Re-usable parts and materials

Dismantling III: Core scrap

Compression

Shredding

Post-shredder processing

Fig. 7.118 Motor vehicle dismantling procedure (Source: BMW)

As a first administrative step, the vehicle is handed over to the dismantling facility, where it is registered and pre-assessed. The former owner then receives a document that certifies legal disposal of the vehicle. Before any part is disassembled, all liquids such as fuels, brake fluids, refrigerants, coolants, all kinds of oils, washer fluids, hydraulic media etc. need to be completely drained and the respective openings closed. Dismantling then proceeds in three main steps: At first, parts containing hazardous or toxic substances (e.g. according to annex II of the EU directive 2000/53/EG) are removed and isolated. Then, parts and components that can be re-used or are transferred to specific recycling facilities (e.g. tires, glass, batteries etc.) are dismantled. At last, core scrap such as engines, gearboxes or axle parts are ripped out of the vehicle. The vehicle is then compressed and transferred to a shredding facility, where it is mechanically shredded into small chips. During the following post shredder processing, material fractions are identified, separated, and transferred to subsequent material or energetic recycling.

Design for Recycling

Recyclability of a vehicle is 100% determined by the selected materials. According to Fig. 7.119, which shows the weight distribution of materials used in the current 7 series BMW, over 70% of a car's materials are metals. As automotive scrap is usually a blend of metal and non-metal materials, it is hackled, sorted and filtered and then further processed in furnaces, where iron, steel, copper and aluminum can be separated in pure form from other metals and added ore.

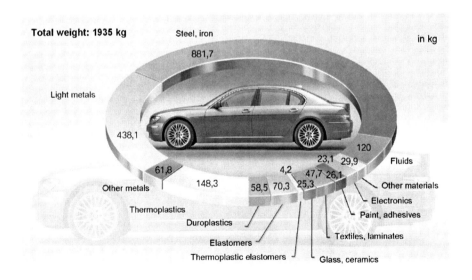

Fig. 7.119 Weight distribution of materials in a motor vehicle (Source: BMW)

Single-variety plastic components can be more or less fully recycled. Granulate from shredded polypropylene (PP) components e.g. is used together with new granulate in the production of especially non-visible automotive parts. Another good example for material recycling is worn tires. The rubber is separated from texture and funicular parts, and then used to produce various goods such as e.g. floor mats for playgrounds, floor sound absorbers for homes and industrial buildings, or running tracks.

Reusability of plastic parts however is more difficult, if different materials are mixed. Polyethylene (PE) and PP e.g. can hardly be separated if mixed. A major share of recycled plastic parts with undefined material mix is burned in furnaces and used as an energy source.

Coolants and brake fluids can be refurbished and used again.

Over the PEP, recycling engineers and designers assess all parts in terms of their recyclability, and document the results in the respective part drawings. According to VDI 2243, design for recycling includes [76]:

- Provision of material marks on parts to allow proper material identification
- Avoidance of environmentally critical materials, e.g. polyvinyl chloride (PVC), mercury, hexavalent chromium)
- Avoidance of composite materials
- Avoidance of recycling-critical materials, which could disrupt specific recycling processes, e.g. transmission oil in the shredder process

Design for Dismantling

Similar to design for maintenance and repair (compare Sects. 8.1.2 and 8.2.2), ease of dismantling is determined by the time and effort needed to open or break joints that were made during component manufacturing or complete vehicle assembly, and the accessibility to the components that are to be removed during the dismantling process. Figure 7.120 shows the different fractions of a front light unit that can be dismantled in less than 30 sec. To ensure that each vehicle achieves the goals set in terms of design for dismantling, virtual and physical disassembly analyses are carried through over the different phases of the PEP.

VDI 2243 lists the following design guidelines that help ensuring easy dismantling [76]:

- Visual perceptibility of parts and components that need to be removed
- Easy access to parts and components that need to be removed
- Application of consistent joining concepts (similar parts have the same connection concept), even over multiple models
- Easy detachability (e.g. usage of clips instead of adhesives)
- Fast and complete drainage of fuels, oils, coolants, brake fluids, battery acid

Fig. 7.120 Single-material variety components of a dismantled headlight (Source: BMW)

To facilitate quick and correct dismantling of vehicles worldwide, the *International Dismantling Information System* (IDIS) was created by a consortium of carmakers to prepare their specific recycling and dismantling oriented product information in a standardized way and make it available to dismantlers worldwide.

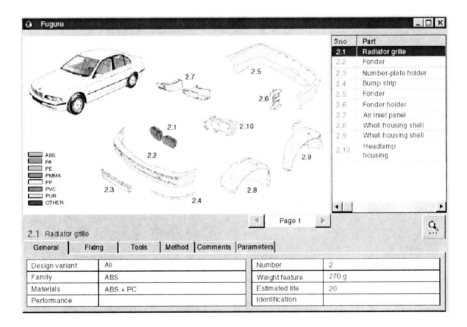

Fig. 7.121 International dismantling information system (Source: VDI [76])

In addition to information relating to recyclable components (e.g. material, weight, estimated dismantling time, recommended dismantling technique etc.), it contains guidelines for handling pyrotechnic components, removal of fluids, or treatment of hazardous or toxic substances. Figure 7.121 shows a sample IDIS user interface.

7.10.6.3 System Integration and Validation

Legislation and society require objective data concerning how well the intended end-of-life procedures for a given vehicle meet the requirements of sustainable development. But while formerly treatment of end-of-life vehicles was mainly determined by ecological considerations, the dramatic increase of material prices as e.g. the steel price over the last years have made easy dismantling and comprehensive recycling also economically interesting for manufacturers. An assessment of dismantling and recycling processes must hence include both monetary and environmental aspects.

- Monetary aspects: Dismantling costs, resale value of refurbished or recycled materials or components
- Environmental aspects: Emissions generated during dismantling, recycling and disposal processes; environmental impact caused by residue that have to be disposed in landfills.

In terms of its ecological and economical value, treatment of vehicle components as part of the end -of-life procedures can be divided into three different levels (from better to worse): Re-use, recycling, and disposal (see Fig. 7.122).

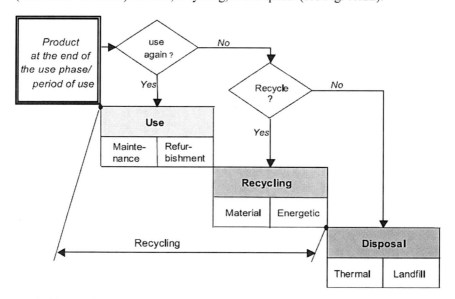

Fig. 7.122 Recycling cascade (Source: VDI [76])

One of the first recyclability assessments has been set up by the USCAR Vehicle Recycling Partnership (VRP). It follows four steps [77]:

1. Identification of components and materials to be included in the assessment
2. Rating of the components recyclability using a rating scheme (see Table 7.17)
3. Calculation of the specific recyclability (recyclability per weight)
4. Identification of improvement potentials

Table 7.17 Recyclability rating scheme (Source: USCAR [77])

Score	Material recyclability	Ease of dismantling
1	Reusable	Can be disassembled easily, manually.
2	Recyclable, infrastructure and technology clearly defined	Can be disassembled with effort, manually.
3	Recycling technically feasible, infrastructure not available	Can be disassembled with effort, requires some mechanical separation or shredding to separate component materials and parts. Process fully proven.
4	Recycling technically feasible, further process or material development required	Can be disassembled with effort, requires some mechanical separation or shredding to separate component materials and parts. Process under development
5	Organic material for energy recovery that cannot be recycled	Cannot be disassembled.
6	Inorganic material, no known technology for recycling	

7.10.7 Pre-usage Sustainability

7.10.7.1 Legal and Customer Requirements

For a realistic eco-balance, considering the environmental impact a vehicle has while being utilized by the customer it is definitely not enough. What really counts in terms of e.g. emissions or energy consumption is the total over the complete life cycle, including development, logistics/transport, and manufacturing - down to the last link in the supply chain (compare Sect. 7.10.1). Not only legislation, but especially customers are getting more and more sensitive to how manufacturers treat the environment and the people involved in the respective processes. Automotive industries need to become – and partially already have become – responsible partners of society as a whole rather than just being "good neighbors".

As reduction of energy consumption directly correlates with reducing costs, manufacturers are usually quick to implement these measures. A good example for the advanced measures OEMs take to reduce energy consumption is the landfill gas project at BMW's plant in Spartanburg SC, where methane gas generated at a landfill nearby is recycled and used to fuel gas turbines that generate electricity for the factory.

Limits for pollution from manufacturing processes are typically strictly controlled by local legislation. Especially critical are chemical processes for painting, coating and cleaning of surfaces, but also e.g. disposal of prototype parts that do not adhere to series parts' recycling requirements.

An aspect often neglected in eco-balances e.g. for alternative powertrain technologies is transport. Every truck, ship, plane, or train that carries parts and components from one factory to another – or complete vehicles from an assembly plant to the customer – generates emissions and consumes energy. A hybrid car e.g. that has its technology critical parts (such as batteries) shipped all over the globe might have posed more burden on the environment prior to being handed over to the customer than its conventionally powered equivalent over its first couple of 10,000 miles of use.

Sustainable management of business also requires maintaining the financial, physical and psychical vigor of the people involved in the business. This includes e.g. health care for employees, ergonomic workstations, and job security. Especially when it comes to low-cost-country sourcing, this important aspect of sustainability requires that manufacturers make sure that cost benefits are not achieved at the expense of human working conditions, or higher pollution.

Last but not least, sustainable development also requires sustainable financial stability as the main pre-condition for a business to survive and thus offer jobs on a long-term scale.

In most countries, applicable law regulates working conditions (especially in terms of payment, working hours, safety) and the pollution caused by production processes. But as the actual requirements of these laws – and how strictly adherence is enforced by authorities – varies greatly from region to region, national and international organizations have formulated sustainability related targets (that are published e.g. as charters, codes of conduct etc.) to which responsible vehicle manufacturers voluntarily commit themselves. Prominent examples of such declarations are:

- *International Chamber of Commerce (ICC) Business Charter for Sustainable Development:*
  The objective of the ICC charter is "that the widest range of enterprises commit themselves to improving their environmental performance in accordance with the principles, to having in place management practices to effect such improvement, to measuring their progress, and to reporting this progress as appropriate, internally and externally" [78].
- *Organization for Economic Cooperation and Development (OECD) Guidelines for Multinational Enterprises:*
  Voluntary principles and standards for responsible business conduct with the following objectives: Ensuring that operations of these enterprises are in harmony with government policies, strengthen the basis of mutual confidence

between enterprises and the societies in which they operate, helping to improve the foreign investment climate, and enhancing the contribution to sustainable development made by multinational enterprises [79].

- *United Nations Environment Programme (UNEP) Cleaner Production Programme:*
  Continuous application of preventative environmental strategies to products, processes, and services with the objective to increase eco-efficiency and to reduce risks to humans and the environment [61].
- *UN Global Compact:*
  A strategic policy initiative for businesses that are committed to aligning their operations and strategies with 10 principles in the areas of human rights, labor, environment and anti-corruption [80].
- *Econsense:*
  An association of 25 globally-active German companies and business organizations who have integrated the principles of sustainable development into their corporate strategies. Econsense's objectives include positive and proactive involvement in the political and societal debate and decision making processes, exchange of experiences and joint development of positions, communication of the problem-solving competence of the business community, clarification of opportunities and limits of corporate responsibility [81].

By adhering to the respective targets, companies show commitment to sustainable development far beyond the legal requirements. But this social and ecological conscientiousness is not only a prerequisite for sustainable economic success. With their increasing tendency to ethical consumerism, customers see such corporate social responsibility not only as a characteristic of the manufacturer but also a characteristic of the product they are willing to pay for.

7.10.7.2 Component and System Design

The possibilities to influence or steer corporate sustainability – as opposed to product sustainability – through product development processes are very limited. But even though the environmental impact of product development is much smaller than the one of production processes, the principles of sustainability can nevertheless be fully applied. Typical aspects of sustainability in product development processes are:

- Working conditions for employees
- Fair selection and treatment of suppliers and service providers
- Preference of suppliers and service providers that for their part adhere to sustainability standards
- Protection of copyrights
- Preservation of energy

- Reduction of hardware prototypes
- Adherence to technical standards

The bigger potential however lies in the vehicle design as a determining factor for sustainability of subsequent processes in production, sales and service. "Design for corporate sustainability" includes aspects such as

- Design for manufacture allows ergonomic and safe work stations especially in assembly.
- Design for service reduces resources needed for maintenance and repair.
- Design for durability increases the recyclability of the vehicle at the end of its life.
- Design for logistics reduces energy consumption and emissions caused by transport of components or complete vehicles.
- Paint and coating without toxic solvents reduce emissions in the paint shop.
- Reduced quality requirements e.g. for surfaces not visible to the customer reduce energy consumption and emissions caused by rework.

7.10.7.3 System Integration and Validation

As discussed above, there are numerous sets of corporate sustainability targets to which carmakers voluntarily adhere. To allow objective validation and comparison of an organizations economic, environmental, and social performance, the *Global Reporting Initiative* (GRI) has developed the so-called Sustainability Reporting Guidelines which represent the world's most widely used sustainability reporting framework [82]. The three main elements are

- Strategy, profile and governance
- Management approach
- Performance indicators: Economic, environmental, labor practices and decent work, human rights, society, product responsibility

The reporting practices can be applied both to the whole organization and to development. It creates transparency and accountability; and at the same time enables performance improvement.

There is however no standardized approach to assess "design for corporate sustainability".

References

1. Albano LD, Suh NP (1994) Axiomatic design and concurrent engineering: Tools for product introduction. Computer-Aided Design 26:499–504
2. UNECE (2002) World Forum for Harmonization of Vehicle Regulations. How it works. How to join in. United Nations, New York

3.   USDOT (1999) Federal motor vehicle safety standards and regulations.
     http://www.nhtsa.dot.gov/cars/rules/import/FMVSS/. Accessed 20 November 2008
4.   Braess HH, Seiffert U (2005) Handbook of automotive engineering. SAE, Warrendale (PA)
5.   USDOT (1998) Compliance testing program.
     http://www.nhtsa.dot.gov/cars/testing/comply/Mission/1_ovsc_1.html.
     Accessed 20 November 2008
6.   Reed P (2008) Revealing the hidden costs of car ownership.
     http://www.edmunds.com/advice/buying/articles/59897/article.html.
     Accessed on 20 November 2008
7.   Victoria Transport Policy Institute (2007) Transportation cost and benefit analysis. Vehicle
     costs. http://www.vtpi.org/tca/tca0501.pdf. Accessed on 20 November 2008
8.   AAA (2008) Your driving costs. How much are you really paying to drive? AAA Associa-
     tion Communication, Heathrow (FL)
9.   Perini G (2007) BMW Group. The art of car design. CAR STYLING Special Edition
10. International Ergonomics Association (2000) What is ergonomics?
     http://www.iea.cc/browse.php?contID=what_is_ergonomics.
     Accessed 20 November 2008
11. Theissen M (2007) Innovative Produktentwicklung. Lecture hand-out, Hochschule für
     Technik und Wirtschaft Dresden
12. Porter JM, Porter CS (2002) Occupant accommodation: an ergonomics approach. In:
     Happian-Smith J (ed) An introduction to modern vehicle design. SAE, Warrendale (PA)
13. Genuit K (2004) The sound quality of vehicle interior: a challenge for the NVH-engineers.
     International Journal of Vehicle Noise and Vibration 1:158–168
14. Plack CJ (2005) The sense of hearing. Psychology press, New York
15. Mozaffarin A, Pankoke S, Bersiner F, Cullmann A (2007) MEMOSIK V – Development
     and application of an active, 3-dimensional dummy for the measurement of vibration com-
     fort on vehicle seats. VDI-Berichte. 2002:481–509
16. Bloemhof H (2002) Fahrzeug-Akustik Heute. Fraunhofer Stuttgart, 29.10.2002.
     http://www.sfb374.uni-stuttgart.de/rv_02_03/Fahrzeugakustik_bloemhof.pdf.
     Accessed 20 November 2008
17. Transport Canada (2003) Strategies for reducing driver distraction from in-vehicle telemat-
     ics devices: A discussion document.
     http://www.tc.gc.ca/roadsafety/tp/tp14133/pdf/tp14133e.pdf.
     Accessed 20 November 2008
18. Strayer DL, Drews FA, Crouch DJ (2006) A comparison of the cell phone driver and the
     drunk driver. Human Factors 2006; 48(2):381–91
19. CE4A (2008) Welcome to CE4A – consumer electronics for automotive.
     http://www.ce4a.org. Accessed 20 November 2008
20. Chrysler (2008) Official HEMI website of Chrysler LLC. http://www.hemi.com.
     Accessed 20 November 2008
21. Lechner G, Naunheimer H (1999) Automotive transmissions: Fundamentals, selection, de-
     sign and application. Springer, Berlin
22. Sommer D, Schmid M, Rohracher J (2007) Eigenschaften von Allradfahrzeugen. ATZ
     109:280-287
23. Harnett P (2002) Objective methods for the assessment of passenger car steering quality.
     VDI Series Nr. 506. VDI, Düsseldorf
24. Breitfeld C, Streng, P (2007) Konzepte für zukünftige Getriebeentwicklungen. ATZ
     109:490-498
25. Mühlbauer R (2007) Virtuelle Fahrdynamik-Entwicklung am Beispiel des neuen X5. In:
     Virtual Vehicle Creation. 11. Internationale Automobiltechnische Konferenz
26. Matschinsky W (1987) Die Radführungen der Straßenfahrzeuge. TÜV Rheinland, Köln
27. Milliken WF, Milliken DL, Olley M (2002) Chassis design: Principles and analysis. SAE,
     Warrendale (PA)
28. Gillespie TD (1992) Fundamentals of race car dynamics. SAE, Warrendale (PA)

29. Hall B (2002) Suspension systems and components. In: Happian-Smith J (ed) An Introduction to modern vehicle design. SAE, Warrendale (PA)
30. Graham A (2003) Steer-by-wire advances at ZF. Automotive News Europe 14 July 2003
31. Becker K, Stickel Th (2006) Einfluss von Karosseriesteifigkeiten auf die Fahreigenschaften eines PKW. In Becker K (ed) Subjektive Fahreindrücke sichtbar machen. Expert, Renningen
32. European Vehicle passive safety network (2004) Roadmap of future automotive passive safety technology development. European Commission 5th Framework Programme
33. Euro NCAP (2008) Test procedures explained.
    http://www.Euro NCAP.com/testprocedures.aspx. Accessed 20 November 2008
34. Euro NCAP (2008) European new car assessment programme. Assessment protocol and biomechanical limits. Version 4.2.
    http://www.Euro NCAP.com/cache/file/5ab781cd-4e7c-4542-b6f7-1d46733289ef/Euro-NCAP-Assessment-Protocol-Version-4.2-June-2008.pdf. Accessed 20 November 2008
35. Zeidler F, Pletschen B, Scheunert D (1993) Development of a new injury cost scale. Accident Analysis & Prevention 25(6):675–687
36. ETSC (2003) Cost effective EU transport safety measures.
    http://www.etsc.be/documents/costeff.pdf. Accessed on 20 November 2008
37. Chinn B (2002) Crashworthiness and its influence on vehicle design. In: Happian-Smith J (ed) An introduction to modern vehicle design. SAE, Warrendale (PA)
38. Ichikawa M, Nakahara S, Wakai S (2002) Mortality of front-seat occupants attributable to unbelted rear-seat passengers in car crashes. The Lancet 359(9300):43–44
39. Glassbrenner D, Jianqiang Y (2006) Seat belt use in 2006 – overall results. NHTSA traffic safety facts.
    http://www-nrd.nhtsa.dot.gov/pdf/nrd-30/NCSA/RNotes/2006/810677.pdf.
    Accessed 20 November 2008
40. NHTSA (2008) THOR advanced crash test dummy.
    http://www-nrd.nhtsa.dot.gov/departments/nrd-51/THORAdv/THORAdv.htm.
    Accessed 20 November 2008
41. FBI (2008) Uniform crime reports. Motor vehicle theft
    http://www.fbi.gov/ucr/cius2007/offenses/property_crime/motor_vehicle_theft.html. Accessed 20 November 2008
42. National motor vehicle theft reduction council (2008) Carsafe. Driving down vehicle theft.
    http://www.carsafe.com.au/t_09.html. Accessed on 20 November 2008
43. FMVSS 114 (1997) Theft protection
44. CMVSS 114 (2005) Locking and immobilization systems
45. ECE R 116 (2006) Uniform technical prescriptions concerning the protection of motor vehicles against unauthorized use
46. Gardner M, Hartley R (2002) Stolen vehicle tracking. ACPO and Home Office guidance to companies on police policy. Home Office scientific development branch, St Albans
47. Thatcham (2006) New vehicle security assessment.
    http://www.thatcham.org/nvsr/. Accessed on 20 November 2008
48. ADAC (2007) Die ADAC Pannenstatistik 2007.
    http://www.adac.de/Auto_Motorrad/pannenstatistik_maengelforum/
    Pannenstatistik_2007. Accessed 20 November 2008
49. VDI Guideline 4001-2 (2006) Reliability terminology
50. IEEE 610.12 (1990) IEEE standard glossary of software engineering terminology
51. Pan J (1999) Software reliability. Dependable embedded systems. Carnegie Mellon University. http://www.ece.cmu.edu/~koopman/des_s99/sw_reliability/index.html.
    Accessed 20 November 2008
52. Barkai J (2004) Quality improvement and warranty cost containment. SAE Transactions 113:392–396
53. Keiller PA, Miller DR (1991) On the use and the performance of software reliability growth models. Reliability Engineering & Systems Safety 32:95–117

54. Rosenberg L, Hammer T, Shaw J (1998) Software Metrics and Reliability. NASA SATC. http://satc.gsfc.nasa.gov/support/ISSRE_NOV98/software_metrics_and_reliability.html. Accessed 20 November 2008
55. Hubing T (2008) Introduction to EMC. The Clemson University Vehicular Electronics Laboratory. http://www.cvel.clemson.edu/emc/tutorials/Introduction_to_EMC/Introduction.html. Accessed 20 November 2008
56. DIN ISO 11452-5 (2003) Road vehicles – Component test methods for electrical disturbance by narrowband radiated electro-magnetic energy. Part 5: Stripline
57. Hubing T (2008) EMC Design Guideline Collection. The Clemson University Vehicular Electronics Laboratory. http://www.cvel.clemson.edu/emc/tutorials/guidelines.html. Accessed 20 November 2008
58. VDA Volume 4 (2003) Quality assurance prior to serial application. Quality assurance during product realization. Methods and procedures
59. NASA (2002) Fault tree handbook with aerospace applications. NASA office of safety and mission assurance. NASA, Washington (DC)
60. IEC 61025 (2006) Fault tree analysis (FTA)
61. UNEP (2008) Understanding cleaner production. http://www.uneptie.org/scp/cp/understanding/concept.htm#1. Accessed 20 November 2008
62. Shell (2008) Shell energy scenarios to 2050. Shell International, The Hague
63. VDA (2008) Auto Annual Report 2008
64. CARB (2008) Fact sheet. California vehicle emissions. http://www.arb.ca.gov/msprog/zevprog/factsheets/calemissions.pdf. Accessed 20 November 2008
65. EPA (2008) Fuel economy label. http://www.epa.gov/fueleconomy/label.htm. Accessed 20 November 2008
66. CARB (2008) Fact sheet. The environmental performance label - coming soon to California vehicles. http://www.arb.ca.gov/msprog/labeling/eplabelfs.pdf. Accessed 20 November 2008
67. Tesla (2008) Tesla Motors. Performance. http://www.teslamotors.com. Accessed 20 November 2008
68. Helmolt R, Eberle U (2008) Fuel cell vehicles: Fundamentals, system efficiencies, technology development, and demonstration projects. In: Léon A (ed) Hydrogen technology. Mobile and portable applications. Springer, Berlin
69. 70/220/EEC (1970) Measures to be taken against air pollution by emissions from motor vehicles. Annex III, Appendix 1
70. Jones CW, Friedmann S, Andre M et al. (1998) HYZEM: A joint approach towards understanding hybrid vehicle introduction into Europe. IMECHE Conference Transactions 1998:3–15
71. Den Boer LC, Schroten A (2007) Traffic noise reduction in Europe. Health effects, social costs and technical and policy options to reduce road and rail traffic noise. CE, Delft
72. ISO 362 (2007) Measurement of noise emitted by accelerating road vehicles. Engineering method. Part 1: M and N categories
73. BMW GS95002 (2004) Electro-magnetic compatibility. Requirements and tests. BMW group standard
74. 2000/53/EC (2000) End-of-life vehicles
75. EPA (2008) National vehicle mercury switch recovery program. http://www.epa.gov/mercury/switch.htm. Accessed 20 November 2008
76. VDI 2243 (2002) Recycling-oriented product development
77. Coulter S, Bras B, Winslow G, Yester S (1996) Designing for material separation: Lessons from automotive recycling. Journal of Mechanical Design 120(3):501–509
78. BSDglobal (2008) ICC Business Charter for Sustainable Development. http://www.bsdglobal.com/tools/principles_icc.asp. Accessed 20 November 2008

79. OECD (2000) Guidelines for multinational enterprises.
    http://www.oecd.org/department/0,3355,en_2649_34889_1_1_1_1_1,00.html.
    Accessed 20 November 2008
80. UN (2008) Overview of the UN Global Compact.
    http://www.unglobalcompact.org/AboutTheGC/index.html.
    Accessed 20 November 2008
81. Econsense (2008) Dialogue platform and think tank for CSR.
    http://www.econsense.de/_ENGLISH/index.asp. Accessed 20 November 2008
82. GRI (2006) G3 guidelines.
    http://www.globalreporting.org/ReportingFramework/G3Guidelines/.
    Accessed 20 November 2008

Chapter 8
Secondary Complete Vehicle Characteristics

Abstract   In contrast to the direct customer relevant characteristics discussed throughout the previous chapter, secondary characteristics represent requirements of *internal customers* of the development process, namely production and service. Just like real customers, they also have requirements which the vehicle must meet and which consequently must be considered throughout the PEP.

8.1 Production Integration

8.1.1 Legal and Internal Customer Requirements

The first – though internal – customer of development is the plant in which the respective vehicle will be produced. Compared to the creative, project-driven atmosphere of a development center, a plant represents a completely different world: The dream of an automotive production plant's manager is a production process that ticks as precisely and smoothly as a clockwork, creating a constant stream of parts, components and eventually complete vehicles of perfect quality (see Fig. 8.1). Every single part that is delivered to the plant must be available at the right place at the right time – and then exactly fit to its intended location in the vehicle.

How well this production clockwork runs is determined by two main factors:

- The production environment, including equipment functionality and availability, workforce qualification, and process maturity. Responsible for the production environment are the process planners of the respective plant.
- The degree to which the parts' and vehicles' design is in compliance with the requirements of i.e. manufacturing, assembly and logistics. Realization of this complete vehicle characteristic is pursued by production integration engineers (compare Sect. 1.3.2.2) – which in contrast to the process planners belong to development.

J. Weber, *Automotive Development Processes*, DOI 10.1007/978-3-642-01253-2_8,
© Springer-Verlag Berlin Heidelberg 2009

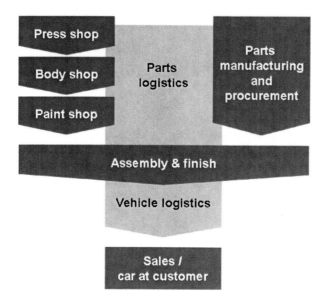

Fig. 8.1 Complete vehicle production process

Against this background, one of the main tasks of production managers is to prevent any disturbance that could disrupt the clockwork. Planned and unplanned changes to the production system must be handled in a way that they do not affect production volume, quality and cost. And while the production environment and related changes lie in their own responsibility, production managers are naturally wary of design changes they can not control. Henry Ford's famous statement according to which he wished that all cars produced would be black e.g. reveals him as a man of production rather than of design. Table 8.1 provides examples for the different types of process disturbances.

Table 8.1 Classification of production process disturbances

	Production environment	**Vehicle design**
Planned	• Equipment maintenance	• Design changes
	• Installation of new equipment	• Model year measures
	• Induction of new workforce	• Launch of new variants and/or options
	• Continuous process improvements	• Launch of new model
Unplanned	• Repair	• Parts quality issues
	• Parts shortage	
	• Non-conformance of parts	
	• Absence of trained workforce	

The biggest challenge in this context is the launch of a new model. In order to be able to integrate it as smoothly as possible into the production environment, production poses tight requirements regarding a vehicle's design [1]. To be able to easily produce painted bodies or assemble complete vehicles in perfect quality a couple of hundred times a day, parts and components must be designed especially for:

- Logistics: Easy identification, usability of existing transport equipment, prevention of transport damages, minimization of required transport space etc.
- Fitting: Provision of obvious and unambiguous fitting paths
- Fixing: Tool accessibility, direct visual or acoustic feedback confirming correct fitment etc.
- Joining: Facilitate joints (bolts, clips, rivets, welding, adhesives etc.) by provision of the respective part geometries
- Safety and ergonomics: Alignment of a parts weight and shape with manual handling requirements; prevention of possible injury, prevention of contact with toxic or noxious substances etc.

Parts of these requirements have already been discussed in Sect. 7.10.7 as design requirements that determine sustainability of production processes. Legal requirements only exist with regards to safety and ergonomics.

8.1.2 Component and System Design

Design for production integrates three subordinated principles: Design for manufacturing, design for assembly, and design for logistics. On a single part's level, the part's geometry, material, and respective manufacturing technology determine quality and cost of manufacturing (e.g. stamping of sheet metal body parts or injection molding of polyamide air intakes). A detailed collection of respective design guidelines can be found e.g. in [2].

On a sub-assembly or complete vehicle level, the determining factors for quality, cost and time are parts handling, joining, and fastening. Design for manufacturing and assembly is not only supported by design guidelines (as e.g. shown in [3]), but also by elaborate methodologies and software-tools that focus on the reduction of costs. A prominent example is *Design For Manufacturing and Assembly* (DFMA) [4].

Over the PEP, production integration engineers consistently check the vehicle's compliance with the production requirements using both virtual and physical mock-ups.

Because of their flexibility and cost-efficiency, virtual mock-ups are the preferred means for assessing assembly processes, not only during concept phase. Container, fixtures, tools and man-models can be integrated into the virtual scenes – which allows a realistic evaluation of transport and handling, fixing processes, and ergonomics [5]. As an example, Fig. 8.2 shows a typical virtual scene, in which a man model assembles a part using a power tool. The comprehensive analysis of this virtual scene allows e.g. the evaluation of the accessibility of the bolt with the specified power tool, the worker's field of vision as a prerequisite for proper fitment and fixing, and – by analysis of the man model's movement – the ergonomics of the workplace.

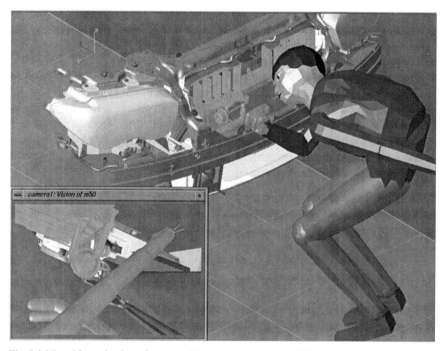

Fig. 8.2 Virtual investigation of an assembly process (Source: BMW)

While virtual vehicle mock-ups represent a geometrically ideal world of rigid components, physical mock-ups allow investigations that include real-world physical effects such as plastic or elastic material deformation or production tolerances. At the beginning of each prototype build phase, laminated bodies and other rapid-prototyping techniques are used to investigate the behavior of deformable, elastic or labile parts prior to the actual build. Figure 8.3 shows the respective timeframe.

Figure 8.4 shows such a laminate mock-up that is used to assess routing of the main harness in the front part of the cabin. The results from this investigation can still be allowed for in the prototype build process.

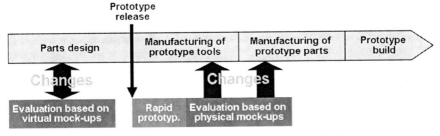

Fig. 8.3 Timeframe for production evaluation as part of a prototype build phase

Fig. 8.4 Assessment of harness routing using a laminate mock-up (Source: BMW)

8.1.3 System Integration and Validation

Before responsibility in a vehicle project is handed over from development to production, and the first pre-series cars are built in the plant, the plant as the internal customer signs off on the complete production process. While the fitness for assembly of separate parts can be easily evaluated as part of the design process, the complete vehicle production sign-off requires a comprehensive assessment of all production process steps in the correct order and according to different criteria in a so-called process week. Here, part by part is added to the vehicle and evaluated according to 11 criteria by experts from design and production, exactly following the specified assembly sequence [6]. For a vehicle with several options and engine variants, this can take up to two weeks of intensive work. Figure 8.5 shows the evaluation matrix in which the respective criteria for each assembly step are rated green, yellow, or red.

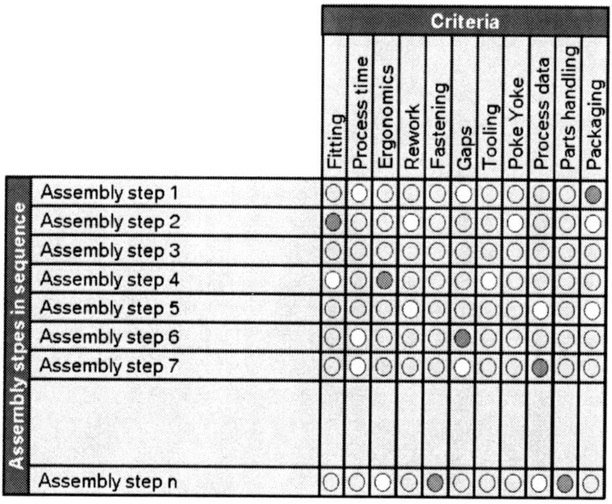

Assembly steps in sequence	Fitting	Process time	Ergonomics	Rework	Fastening	Gaps	Tooling	Poke Yoke	Process data	Parts handling	Packaging
Assembly step 1	○	○	○	○	○	○	○	○	○	○	●
Assembly step 2	●	○	○	○	○	○	○	○	○	○	○
Assembly step 3	○	○	○	○	○	○	○	○	○	○	○
Assembly step 4	○	○	●	○	○	○	○	○	○	○	○
Assembly step 5	○	○	○	○	○	○	○	○	○	○	○
Assembly step 6	○	○	○	○	○	●	○	○	○	○	○
Assembly step 7	○	○	○	○	○	○	○	○	●	○	○
Assembly step n	○	○	○	○	●	○	○	○	○	●	○

Fig. 8.5 Process evaluation matrix

To allow objective rating of the process steps regarding to each criteria, exact specifications are given concerning when a criterion should be rated as green, yellow or red. Table 8.2 shows the specifications taking the criteria "fitting" and "process time/sequence" as an example.

To facilitate prototype build and to boost process maturity, *process week* assessments are also carried through at the beginning of the respective prototype build phase: One process week using virtual vehicles before the respective prototype parts are ordered, and one real process week using the parts that have been delivered for the first prototype. Figure 8.6 shows a scene from a virtual process week: The left projector shows a section of the virtual vehicle, where part after part is added according to the assembly sequence; the right projector shows the evaluation and documentation system.

Table 8.2 Process rating scheme (Source: BMW)

Criterion	Rating	Specification
Fitting	Red	Part cannot be installed, high damage risk, part dimensions conflict with the available mounting locations
	Yellow	Requires high assembly forces, mounting location not clearly identifiable (optics, audio, haptics), accessibility only partially guaranteed
	Green	Assembly process-stability achievable, installation space and assembly path secured.
Process time / sequence	Red	The assembly cannot be completed in the allotted time/the part cannot be assembled in the required sequence, structural changes required
	Yellow	The assembly time considerably exceeds the specification/the part cannot be assembled in the planned sequence, adjusted cycle time necessary
	Green	The assembly time is within specification / the assembly sequence is achievable

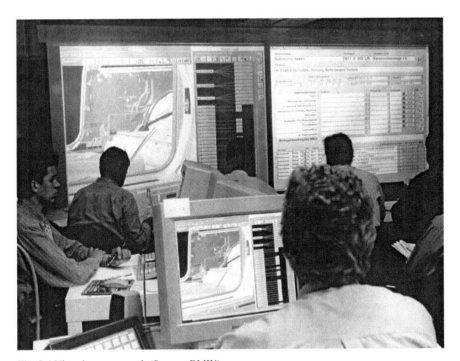

Fig. 8.6 Virtual process week (Source: BMW)

8.2 Service Integration

8.2.1 Legal and Internal Customer Requirements

After a vehicle has been handed over to the customer, the next business unit of the OEM that deals with it is service. Another internal customer, service performs planned maintenance, upgrades or repair. From their point of view, a vehicle's design must allow for simple, fast and inexpensive service processes, which include:

- Maintenance and upgrade:
  - Vehicle check
  - Exchange of wearing parts and consumables
  - Installation of additional components (e.g. aftersales parts)

- Repair:
  - Problem diagnosis
  - Removal of damaged parts
  - Assembly of new parts

Service requirements include minimum maintenance needs (see also Sect. 7.9.1 for reliability and Sect. 7.2.1 for maintenance costs) and for the most part also go in line with requirements for easy dismantling (see Sect. 7.10.6.1). They are – as service processes are mostly disassembly and re-assembly processes – at first glance even equal to production's requirements (see Sect. 8.1.1). There are however significant differences:

- In contrast to a plant that builds one model at a time, service has to be able to deal with all current and former models of the brand the respective service station represents. A customer with a 20 year old car expects to get the same kind of service he would have with a brand new vehicle.
- While in a plant one worker assembles one set of components again and again using highly optimized tools and equipment, service staff has to fulfill much broader work content with generic equipment and a standardized set of tools, often dealing with several different vehicles in parallel. This includes IT tools for "repair" or upgrade of product software.
- Prior to repair, service staff has to diagnose the customer's problem.
- Service staff has to handle physically damaged parts.

8.2.2 Component and System Design

Just as a production plant is represented in a vehicle development project by production integration engineers (compare Sect. 1.3.2.2) who ensure manufacturability of the vehicle under series conditions, service has its objectives represented by service integration engineers. Their tasks over the PEP include:

- Deployment of complete vehicle service concepts that especially allow for simple and fast diagnosis and accessibility for maintenance and repair
- Calculation, planning and evaluation of the respective maintenance, update and repair processes, including service tools (hardware and IT) and spare parts logistics
- Vehicle specific training of service staff
- Support of aftersales components development

Design for maintenance and repair starts with a vehicle architecture that allows fast diagnosis. Following the principles of axiomatic design, this requires a clear and unambiguous relation between parts and functions (compare Sect. 3.1.1). Electronic problems e.g. can be analyzed much easier in a bus structure that provides separate branches for related functions. Also, on-board diagnostic systems allow fast and target-oriented trouble shooting.

After diagnosis has revealed or confirmed which parts of the vehicle have to be exchanged, design for maintenance and repair incorporates easy access to these parts. As – in contrast to series production, where parts are attached to an incomplete vehicle that becomes more and more complete with every part that is added - service operations always start from a complete vehicle, the respective translation path for disassembly and re-assembly of these parts in service is usually different to the one used in production. Here, an obvious example is the engine, which in series production is typically mounted to the vehicle from below as part of the preassembled powertrain and chassis in a highly automated marriage station. However, to exchange the engine in service this process can not simply be reversed. Instead, the engine is released from the gearbox and then removed upwards through the engine compartment (see Fig. 8.7).

The specified set of service tools (that must be available in all relevant service stations) should be accepted by vehicle design as a rigid framework that should only be violated if absolutely necessary. While in series production tools can easily be (and are) individually optimized for each fixing task, a design solution that requires a new service tool initiates significant costs and effort by making it necessary to introduce the tools in all service stations worldwide, including training, maintenance etc.

As service staff has to handle all models, one of the objectives of design for maintenance and repair is to apply equal service concepts to as many models as possible. Especially standard operations like changing an air filter or replacing a damaged bumper should be similar (in terms of sequence, tools, transfer path etc.) for each model, so that service staff can improve speed and quality of work by developing a certain routine. Guidelines regarding design for maintenance and repair can e.g. be found in [7] and [8].

Fig. 8.7 Virtual assessment of the engine removal process (Source: BMW)

On-board diagnostic systems in connection with on-board telecommunication systems allow a completely new way of maintenance and repair. With BMW Teleservice e.g., vehicles regularly analyze the status of their wearing parts and send them to the owner's service partner. There, the data is used e.g. to determine and propose the technically appropriate point in time for the next service appointment. If a service appointment is fixed, all vehicle data the service partner needs for preparing the visit are automatically transferred. Also, malfunctions can be remotely diagnosed and – especially electronic problems – even be fixed without taking the vehicle to the workshop.

8.2.3 System Integration and Validation

To evaluate disassembly and re-assembly of parts in service, basically the same tools and methods are applied as for production: Virtual and physical mock-ups, process weeks etc. (compare Sects. 8.1.2 and 8.2.3). The criteria used to rate the processes however are adapted to the technical conditions of a service environment. As an example, complex manual operations that require careful and dextrous handling of parts (such as cautiously pushing away plastic parts in order to access a bolt) or are ergonomically unfavorable (such as over-head operations) are tolerable in service while being unacceptable in series production.

References

1. Weber J (1999) Optimierung des Serienanlaufs in der Automobilproduktion. VDI-Z 141(11/12):23–25
2. Bralla JG (1998) Design for manufacturability handbook. McGraw-Hill, New York
3. Bramble K (2007) Engineering design for manufacturing. Engineers edge, Monroe (GA)
4. Boothroyd G, Dewhurst P, Knight W (2008) Product Design for manufacture and assembly. Marcel Dekker, New York
5. Weber J (1998) Ein Ansatz zur Bewertung von Entwicklungsergebnissen in virtuellen Szenarien. Institut für Werkzeugmaschinen und Betriebstechnik, Universität Karlsruhe
6. Weber J (2001) Den Serienanlauf sicher im Griff - ein erfahrungsbasiertes Kennzahlensystem zur Absicherung der Anlaufvorbereitung im Automobilbau. VDI-Z 143(1):76–78
7. Shetty D (2002) Design for product success. Society of manufacturing engineers. Dearborn (MI)
8. Anderson D M (2008) Design for manufacturability & concurrent engineering. CIM Press. Cambria (CA)

Abbreviations

A

AAAM	Association for the Advancement of Automotive Medicine
AACN	Advanced Automatic Crash Notification
AAMA	American Automobile Manufacturers Association
ABS	Anti-lock Braking System
ABS	Anti-blockage Brake System
ACEA	Association des Constructeurs Européens d'Automobiles (European Automobile Manufacturers Association)
ADAC	Allgemeiner Deutscher Automobilclub (General German Automobile Association)
AFS	Adaptive Front Steering
AIA	American Insurance Association
AIS	Abbreviated Injury Scale
ALR	Automatic Locking Retractor
APEAL	Automotive Performance Execution and Layout
API	Application Programming Interface
ASC	Automatic Stability Control
ASQ	American Society for Quality
ASTM	American Society for Testing and Materials
AT PZEV	Advanced Technology Partial Zero Emission Vehicle
ATD	Anthropomorphic Test Device
ATSVR	After Theft Systems for Vehicle Recovery
AUTOSAR	Automotive Open System Architecture

B

BA	Brake Assistant
BioRID	Biofidelic Rear Impact Dummy
BOM	Bill of Materials

C

CAD	Computer Aided Design
CAE	Computer Aided Engineering
CAFE	Corporate Average Fuel Economy
CAN	Car Access Network
CARB	California Air Resources Board
CAS	Car Access System
CBC	Cornering Brake Control
CCB	Change Control Board
CE	Consumer Electronics
CE4A	Consumer Electronics for Automotive
CEO	Chief Executive Officer
CFR	Code of Federal Regulations

J. Weber, *Automotive Development Processes*, DOI 10.1007/978-3-642-01253-2_BM2,
© Springer-Verlag Berlin Heidelberg 2009

CFR	Constant Failure Rate
CMMI	Capability Maturity Model Integration
CMVSS	Canada Motor Vehicle Safety Standard
CoC	Center of Competence
COF	Coefficient of Friction
CPU	Central Processing Unit
CRABI	Child Restraint Air Bag Interaction
CSG	Crankshaft starter generator
CSI	Customer Satisfaction Index
CVT	Continuously Variable Transmission

D

DC	Direct Current
DFMA	Design For Manufacturing and Assembly
DIN	German Institute for Standardization (Deutsches Institut für Normung)
DJSI	Dow Jones Sustainability Index
DME	Digital Motor Electronics
DOD	Department of Defense
DSC	Dynamic Stability Control
DTC	Data Trouble Code

E

E/E	Electrics and Electronics
E/E	Electrical / Electronic
EBD	Electronic Brake Force Distribution
ECR:	Engineering Change Request
ECSS	European Customer Satisfaction Survey
ECU	Electronic Control Unit
ECWVTA	EC Whole Vehicle Type Approval
EEPROM	Electronically Erasable Read-only Memory
EFQM	
ELR	Emergency Locking Retractor
ELV	End-of-life Vehicles
ELVS	End Of Life Vehicle Solutions Corporation
EMC	Electro-magnetic Compatibility
EN	European Standard (Europäische Norm)
EOP	End of (Series) Production
EPA	Environmental Protection Agency
ESP	Electronic Stability Program
ETSC	European Transport Safety Council

F

FCEV	Fuel Cell Electric Vehicle
FD	Fault Density
FEM	Finite Element Method
FMEA	Failure Mode and Effects Analysis

FMECA	Failure Mode and Effects Criticality Analysis
FMVSS	Federal Motor Vehicle Safety Standard
FTA	Fault Tree Analysis

G

GDV	Gesamtverband der Deutschen Versicherungswirtschaft e.V.
	(German Insurance Association)
GMR	Giermomentenregelung
	(Yaw Moment Control)
GRI	Global Reporting Initiative
GSM	Global System for Mobile Communications
GTR	Global Technical Regulation

H

HEV	Hybrid Electrical Vehicles
HIL	Hardware-in-the-loop
HMI	Human Machine Interface
HUD	Head-up Display
HVAC	Heating, Ventilating, and Air Conditioning

I

ICC	Integrated Chassis Control
ICC	International Chamber of Commerce
ICS	Injury Cost Scale
IDIS	International Dismantling Information System
IEA	International Ergonomics Association
IEEE	Institute of Electrical and Electronics Engineers
IIHS	Insurance Institute for Highway Safety
INCOSE	International Council of Systems Engineering
IQS	Initial Quality Survey
ISO	International Organization for Standardization
ITC	Inland Transport Committee

J

| JAMA | Japan Automobile Manufacturers Association |

L

| LEV | Low Emission Vehicle |

M

MAIS	Maximum AIS Value
MBS	Multi-body System
MIL	Model-in-the-loop
MIRRC	Motor Insurance Repair Research Centre
	("Thatcham")
MOST	Media Oriented System Transport
MTBF	Mean Time Between Failures
MY	Model Year

N
NCAP   New Car Assessment Programme
NCBS   New Car Buyer Survey
NHTSA   National Highway Traffic Safety Administration
NIST   National Institute of Standards and Technology
NPV   Net Present Value
NVES   New Vehicle Experience Study
NVMSRP   National Vehicle Mercury Switch Recovery Program

O
OECD   Organization for Economic Cooperation and Development
OEM   Original Equipment Manufacturer
OICA   Organisation Internationale des Constructeurs d'Automobiles
            (International Organization of Motor Vehicle Manufacturers)
OSEK   Offene Systeme und deren Schnittstellen für die Elektronik in
            Kraftfahrzeugen
            (Open Systems and their Interfaces for the Electronics in Motor
            Vehicles)

P
PCB   Printed Circuit Board
PDA   Personal Digital Assistant
PDM   Product Data Management
PE   Polyethylene
PEP   Product Evolution Process
PM   Particular Matter
POM   Polyoxymethylene
PP   Polypropylene
PPP   Poly-para-phenylene
PVC   Polyvinyl Chloride
PZEV   Partial Zero Emission Vehicle

Q
QAS   Quality Audit Survey
QFD   Quality Function Deployment
QMS   Quality Management System

R
RAM   Random Access Memory
ROI   Return on Investment
ROM   Read-only Memory
RTE   Run-time Environment

S
SAV   Sports Activity Vehicle
SEI   Software Engineering Institute at Carnegie Mellon University
SHED   Sealed Housing Evaporative Determination
SID   Side Impact Dummy
SIL   Software-in-the-loop

SOP	Start of Production
SRS	Supplemental Restraint System
SSI	Sales Satisfaction Index
StVZO	Straßenverkehrs-Zulassungs-Ordnung
	(Road Traffic Licensing Regulations)
SULEV	Supra Ultra Low Emissions Vehicle
SUV	Sports Utility Vehicle
SWEBOK	Software Engineering Body of Knowledge

T

TCEQ	Texas Commission on Environmental Quality
TCS	Traction Control System
TQM	Total Quality Management
TTCAN	Time-trigged Car Access Network
TÜV	Technischer Überwachungsverein
	(German Technical Inspection Agency)

U

ULEV	Ultra Low Emission Vehicle
UML	Unified Modeling Language
UML-RT	Unified Modeling Language - Real Time
UNECE	United Nations Economic Commission for Europe
UNEP	United Nations Environment Programme
USCAR	United States Council for Automotive Research
USDOD	United States Department of Defense
USDOT	United States Department of Transportation
USP	Unique Selling Proposition

V

VDA	Verband Deutscher Automobilhersteller
	(German Car Makers Association)
VDS	Vehicle Dependability Study
VIN	Vehicle Identification Number
VOC	Volatile Organic Compound
VRP	Vehicle Recycling Partnership

Z

ZEV	Zero Emission Vehicle

Index

Breinigsville, PA USA
12 January 2011
253152BV00005B/19/P